UNIVERSAL SMART GRID AGENT FOR DISTRIBUTED POWER GENERATION MANAGEMENT

By Eric MSP Veith

eveith@veith-m.de

λόγος
LOGOS VERLAG BERLIN

Bibliografische Information der Deutschen Nationalbibliothek

Die Deutsche Nationalbibliothek verzeichnet diese Publikation in der
Deutschen Nationalbibliografie; detaillierte bibliografische Daten sind
im Internet über http://dnb.d-nb.de abrufbar.

Zugl.: Diss., Technische Universität Bergakademie Freiberg, 2017

ISBN 978-3-8325-4512-3

Logos Verlag Berlin GmbH
Comeniushof, Gubener Str. 47,
10243 Berlin
Tel.: +49 (0)30 42 85 10 90
Fax: +49 (0)30 42 85 10 92
INTERNET: http://www.logos-verlag.de

Abstract

Renewable energy sources provide an ever increasing amount of the global electric power generation. However, for many regions, even whole countries, the go-to primary renewable energy sources that are available in large quantities are wind and solar radiation, which are highly volatile since they depend on a factor that is not human-controllable: the weather. The traditional power grid featured centralized power generation and a hierarchical structure; but in addition to the volatility, renewable energy sources blur the distinction between generator and consumer: Through photovoltaic panels on rooftops, a consumer can alternate between becoming a generator and a consumer during any one day. Moreover, wind farms and larger photovoltaic power plants feed into the intermediate layer of the power grid, the distribution grid. Today, lower levels of the power grid feed back into the upper levels, voiding the traditional hierarchical structure of the power grid; power generation has also become distributed, just as the consumption of power. Due to technical restrictions that are inherent in the type of power plant, the traditional coal, oil, or nuclear power plants cannot accompany this volatility with arbitrarily changing their power generation.

At the same time, the notion of the smart grid introduces a vast array of new data coming from sensors in the power grid, at wind farms, power plants, and consumers. The new wealth of information can help in managing the different actors in the power grid. This thesis proposes to view the outlined problem of power generation and distribution as a problem of information distribution and processing.

To accommodate the new, decentralized architecture of the power grid, an equally decentralized approach to grid-wide information processing and distribution is sensible. Each power plant, substation, transformer, and large consumers, such as factories, become agents that exhibit proactive behavior

i

and communicate to maintain the grid-wide power balance.

Local forecasting is the basis for these entities. Every agent forecasts its future power balance or imbalance from historic data. The agents utilize individually trained Artificial Neural Networks to exhibit this forecast. The agent now seeks the help of other agents to solve this disequilibrium. The rules of this exchange are governed by a protocol designed in this thesis. The core principle of the rules that govern the information exchange is to arrive at a power equilibrium while being as scalable as possible without any agent having a priori knowledge of other agents. For this, the protocol remodels the power grid in the communication architecture to take advantage of the properties of the electric grid. Which agent contributes which part to the power equilibrium, however, remains a combinatorial problem. This thesis models the demand and supply of power in the Boolean domain. The power balance solver leverages Ternary Vector Lists and the XBOOLE system to master the emerging complexity. Thus, a distributed demand-supply calculation is defined.

The thesis proves the feasibility of this approach and introduces a metric that combines the information-centric world of the agent software with the world of the power grid. This metric, the data efficiency, shows the impact of the computational approach and enables the comparison of different approaches. This thesis then shows the efficiency gain in terms of line loss avoided based on this metric, comparing the Universal Smart Grid Agent to other solutions.

Acknowledgements

I sincerely thank Prof. Dr. Bernd Steinbach, my doctoral thesis supervisor, from the bottom of my heart for his tremendous amount of help, feedback, and encouragement during the creation of this thesis. Without your trust, I would never have been able to even start this dissertation. Without your patience and your time, I would never have been able to complete it.

I feel deeply indebted to Prof. Dr. Johannes Windeln, my mentor, who enabled me to pursue this goal. Had you not given me this chance, and supported me through these years, I would not be where I am today.

I am grateful to Prof. Dr. Jürgen Otten, my second supervisor, who very kindly supported me, gave me valuable suggestions, and helped me in many ways.

I also specifically thank the German Ministry for Economic Affairs and Energy for funding the comCIGS II project. The members of this project provided me with valuable feedback to my work and ideas.

Ralph Bothe, mayor of the association of municipalities Monsheim, and Willi Bayer and Stefan Radmacher of AöR Energieprojekte Monsheim provided me with data for model development. Uwe Ohl and Sven Wagner of EWR AG, Dr.-Ing. Frank Wirtz of Bayernwerk AG, and Robert Heiliger of E.ON AG helped me to test my ideas against real-world scenarios. I am grateful for their support.

Johannes of Salisbury cites Bernhard of Chatres in his famous quotation, "Dicebat Bernardus Carnotensis nos esse quasi nanos gigantum umeris insidentes, ut possimus plura eis et remotiora videre, non utique proprii visus acumine, aut eminentia corporis, sed quia in altum subvehimur et extollimur magnitudine gigantea," most commonly known in this version by Sir Isaac Newton: "If I have seen further it is by standing on ye shoulders of giants." I am very grateful to Prof. Dr. Martin Ruppert, who sparked my interest in machine learning and,

iii

specifically, Artificial Neural Networks, and who endowed me with a remarkable algorithm. And I cannot even begin to count how many valuable suggestions I received during conferences, at talks or in e-mails from researchers all over the world.

But for a work like this to happen, I also had to lean on the important people in my life, who supported me, listened to my problems, and morally helped me wherever they could: My mother, whose kind words straightened me up whenever I was crestfallen; my father, whose immense support gave me the freedom to write this thesis as it exists today; my grandparents, who always listened to me; and my beloved Sabine, who was always at my side and supported me in uncountable ways. I am deeply thankful that you are a part of my life. I cannot begin to image how I would have been able to face all these challenges had it not been for you.

Contents

List of Figures

List of Tables

List of Algorithms

List of Symbols

Symbol	Meaning
A_i	A particular agent i
$A_{i,t}$	State of agent i at simulation clock t
$\mathbb{B}$	The Boolean space
B	Susceptance, in Siemens
C	Capacity of an information channel, in bit/s
C_i	Constraint definitions for agent i
$\mathrm{C}(\boldsymbol{x_{i,\tilde{t},\tilde{P}}})$	A cover function indicating a power balance equilibrium for the argument vector
$\mathrm{d}(r_i)$	A function that calculates the distance value of the requirement sent by agent i
$\mathrm{d}(x, y)$	A function that calculates the distance or distortion between two values
D	Data volume
D_W	Architecture-dependent word size, in bits
$\frac{D}{D_W}$	Data volume, in words
η	Efficiency, in percent
F_i	Forecaster Module of agent i
G	Conductance, in Siemens
i	A unique agent ID, $i \in I$
I	Set of all unique agent IDs
I'	Set of all unique agent IDs which are explicitly known, e.g., as part of a simulation run description or because they have responded to a request; $I' \subseteq I$
j	A particular message ID
J	Set of all active message IDs

k	General counter variable
k	Boltzman constant ($1.380\,648\,52 \times 10^{-23}\,\mathrm{J\,K^{-1}}$)
L_i	Ordered set of connections (links) of the agent i
l_i	A particular connection (link) of agent i
l	Length, in meters
M_i	Messaging Module of agent i
m_j	A particular message j, $j \in J$
$\mathcal{N}(\mu, \sigma^2)$	A normal distribution with mean μ and variance σ^2
n	Any number of items
O	Set of all objects in the Multipart Evolutionary Algorithm
o	An object, as an individual in a population
$\boldsymbol{o_s}$	Scatter vector of an object
$\boldsymbol{o_p}$	Parameter vector of an object
o_i	An offer from agent i
$\boldsymbol{p}, \boldsymbol{q}$	Any vector
p	Pressure, in bar
P	Real power, in kilowatts
$\tilde{P}$	A power interval, $\tilde{P} = [P_1; P_2]$
P, Q	Any set or Ternary Vector List
P_i	Power balance of agent i
q	A symbol of an alphabet
$q \prec q'$	q precedes q' lexicographically
$q \preceq q'$	q precedes or is equal to q' with regards to lexicographic ordering
Q	Reactive power, in VAr; also: Rate of flow
Q'_C	Capacity of a cable, in $\mu\mathrm{F/m}$
r_i	A requirement from agent i
$\mathrm{r}_i(\boldsymbol{x_{i,\tilde{t},\tilde{P}}})$	Acceptance function for the requirement r_i
R	Set of all requirement equations; also: Resistance, in Ω
s	A state of a Finite State Machine
S	Apparent power, in voltamperes; also: Set of states of a Finite State Machine
S_D	Final simulation state as desired by the simulation run description
S_t	Simulation state at clock t
S_T	Final simulation state as result of the actual simulation run

$\mathrm{S}^n(\boldsymbol{x})$	A symmetric function with $|\boldsymbol{x}|$ arguments and n 1-bits
Σ	An alphabet
ς	A data source
t	A time value, e.g., as simulation clock
T	A time constance; also:
	Simulation clock at the end of a simulation run; also:
	Absolute temperature of a system, in Kelvin; also:
$\tilde{t}$	A time interval, usually $\tilde{t} = [t_1; t_2)$
θ	Angles
ϑ	Temperature, in degrees Celsius
$\mathcal{U}[0; 1)$	A uniform distribution of random numbers in the interval $[0; 1)$
v	Speed, in meters per second
$\mathbf{v}$	Vector of vertices of a tree
w	Energy density, in Wh/kg; also:
	A weight value
w_G	Weight of the implicit gradient information used in the Multipart Evolutionary Algorithm
$\boldsymbol{w}$	Vector of trainable weights of an Artificial Neural Network
$x_{i,\tilde{t},\tilde{P}}$	An atom of a requirement from agent i denoting the time interval $\tilde{t}$ and the power interval $\tilde{P}$
X	Reactance, in Ω
$X \sim \mathcal{U}[0; 1)$	Drawing of a random number from a uniform distribution (notation is used analogous for normal distributions)
$X_k^{\mathcal{U}[0;1)}$	The k-th drawing of a random number from a uniform distribution (notation is used analogous for normal distributions)
$\mathbf{Y}$	Matrix of admittances

List of Acronyms

AC	Alternating Current
ADSL	Asymmetric Digital Subscriber Line
ANN	Artificial Neural Network
API	Application Programing Interface
AS	Autonomous System
ASIC	Application-Specific Integrated Circuit
BDD	Binary Decision Diagram
BDEW	Bundesverband der Energie- und Wasserwirtschaft e.V.
BGP	Border Gateway Protocol
BV	Binary Vector
BVL	Binary Vector List
CA	Certificate Authority
CIGS	Copper Indium Gallium (Di-) Selenide
CIM	Common Information Model
CPU	Central Processing Unit
DC	Direct Current
DDoS	Distributed Denial of Service
DHCPv4	Dynamic Host Configuration Protocol, version 4
DoS	Denial of Service
EGP	Exterior Gateway Protocol
EREC	European Renewable Energy Council
EU	European Union

| EVBDD | Edge-Valued Binary Decision Diagram |
| EVMDD | Edge-Valued Multi-valued Decision Diagram |

FIFO	First In, First Out
FPGA	Field-Programmable Gate Array
FSM	Finite State Machine

GCD	Greatest Common Divisor
GIS	Geospatial Information System
GPGPU	General-Purpose Graphics Processing Unit
GSM	Global System for Mobile Communications

HAWT	Horizontal Axis Wind Turbine
HSDPA	High Speed Downlink Packet Access
HTTP	Hypertext Transfer Protocol

IAEA	International Atomic Energy Agency
ICMP	Internet Control Message Protocol
IDE	Integrated Development Environment
IEC	International Electrotechnical Commission
iff	if and only if
IP	Internet Protocol
IPCC	Intergovernmental Panel on Climate Change
IPsec	Internet Protocol Security
IPv4	Internet Protocol, version 4
IPv6	Internet Protocol, version 6
ISO	International Standards Organization

| JADE | Java Agent Development Framework |
| JSON | JavaScript Object Notation |

LPEP	Lightweight Power Exchange Protocol
LSD	Link State Database
LSTM	Long Short-Term Memory
LTE	Long-Term Evolution

MAE	Mean Absolute Error
MAS	Multi-Agent System
MDD	Multi-valued Decision Diagram

MSE	Mean-Squared Error
NED	Network Description
OCSP	Online Certificate Status Protocol
ODA	Orthogonal Disjunctive/Antivalent
OSGP	Open Smart Grid Protocol
OSI	Open Systems Interconnection
OSPF	Open Shortest Path First
PDF	Probability Density Function
PSO	Particle Swarm Optimization
PV	Photovoltaic
RMSE	Root Mean Squared Error
RNN	Recurrent Neural Network
SA	Simulated Annealing
SESSL	Simulation Experiment Specification via a Scala Layer
SLP	Standard Load Profile
SPSO	Standard Particle Swarm Optimization
SSTP	Scalable and Secure Transport Protocol
STL	Standard Template Library
TAI	Temps Atomique International (en. International Atomic Time)
TCP	Transmission Control Protocol
TEPCO	Tokyo Electric Power Company
TLS	Transport Layer Security
TTL	Time To Live
TV	Ternary Vector
TVL	Ternary Vector List
UDP	User Datagram Protocol
UML	Unified Modelling Language
UMTS	Universal Mobile Telecommunications System
UUID	Universally-Unique Identifier

VoIP	Voice over IP
VPN	Virtual Private Network
WGS	World Geodetic System
WoT	Web of Trust
XML	Extensible Markup Language
ZBDD	Zero-Suppressed Binary Decision Diagram

Glossary

A

active power Work done per unit of time at a load in the power grid

actuator (agent) Any device, hard- and software alike, that provides the agent with a means to influence its environment

after-heat Heat emitted from a nuclear power plant's reactor core after shutdown

agent This is a piece of software that perceives its environment through sensors and acts upon it through actuators. An agent selects an action in order to maximize its performance in regards to its defined goal.

ancillary service Services necessary to maintain reliable operations of the power grid and to facilitate the transmission of power from generator to consumer. This includes reactive power and voltage control, load following, and loss compensation.

apparent power The combination of active and reactive power in an AC circuit, specifically, the magnitude of the vector sum of active and reactive power

area of effect (simulation environment) A closed polygon describing the area in which a particular effect takes place

B

back-propagation of error A training algorithm for Artificial Neural Networks that derives the adjustment of the individual connection weights from the output error that is back-propagated from output to input layer

back-propagation through time A variant of the back-propagation of error training algorithm that is suitable for training Recurrent Neural Networks

base load The minimum amount of electricity demanded during a 24-hour period

Betz limit This number, modelled and proven by Albert Betz, describes the theoretical maximum efficiency of an ideal wind turbine: $\eta = \frac{16}{27} \approx 59.3\%$.

big endian The byte order, or endianness, describes the order in which a digital word is stored or transmitted. Big endian transmits (or stores) the most significant byte first. Little endian stores/transmits the least significant byte first.

black box This term expresses that the object at hand is opaque to the user and its internals are thus unknown and unobservable. Black-box testing observes the tested object's outputs to a known input and tries to infer a model of the test subject from these observations.

black start A start of a power plant without using outside power

block (power plant) Denotes a unit of a steam-based power plant that summarizes all machinery necessary to generate power, such as the boiler or steam generator, the turbine set, etc. Blocks are commonly characterized by their rated power, e.g., 'a 800 MW block.'

C

constant-speed (wind turbine) This term describes a wind turbine design in which the generator is directly connected to the grid. Therefore, the grid's frequency dictates the generator's speed.

context layer In a Recurrent Neural Network, the layer whose neurons feed the results of the previous activation to neurons in another layer

Contract Net Protocol This is the name of a protocol for agent behavior in which agents can award tasks for other agents to solve. The agents form contracts (and, possibly, subcontracts) for the awarded work items.

D

data quality "The state of completeness, validity, consistency, timeliness and accuracy that makes data appropriate for a specific use." (Schultze-Melling, 2010, p. 256)

decay heat Heat generated by a nuclear power plant's reactor core after power generation has been stopped

Dijkstra's Algorithm A path-finding algorithm

discard work Work (in the sense of the unit) that does not contribute active or reactive power to the grid, e.g., when wind turbines are disconnected from the grid when they could feed in.

discrete-event simulation This is a type of simulation that is driven by events. Time can be kept in an abstract manner in the form of *ticks*, where a tick is defined by the occurrence of one or more events, without the need for a relation to a real date and time.

distribution network The high (not highest) voltage part of the grid distributes power regionally, e.g., to metropolitan areas.

divide et impera Being the classic Latin origin of 'divide and conquer,' this term describes an ancient strategy in warfare: In order to win against an superior enemy, its main force must be split into smaller, separate parts which can now be matched by one's own force and can thus be defeated. In computer science, the term is used analogously: A big problem is broken down into smaller, manageable chunks which are solved one-by-one. Thus, a problem that could not be solved in its entirety can be solved piece by piece.

E

epoch This is the point of origin for the Unix timestamp. The epoch is exactly midnight on January 1st, 1970, UTC. The Unix timestamp is a continuous counter of the seconds that have been elapsed since the epoch.

extra-high voltage Typically 220 kV, 380 kV, 500 kV, 700 kV, or 735 kV; can be up to 1500 kV

H

high voltage Typically 60 kV or 110 kV

Hill Climbing An optimization algorithm utilizing local search that is easily trapped in local minima (or maxima, depending on the goal of the optimization)

host This denote an entity on a computer network that can be identified using its network address. A host is compromised of at least one physical interface, but can have more than one.

hydroelectric power plant A type of power plant where water drives a turbine directly to generate power

I

idempotent This word comes from the Latin words *idem*, 'the same,' and *potentia*, 'capability,' and describess the property of a function to achieve the same result, regardless of how often it is executed.

interval map This structure extends the notion of an interval set with mappings. Since the contents of a set must be distinct, an aggregation function typically exists to solve the case of equal or overlapping arguments.

interval set A set consisting of intervals, i.e., a collection of well defined and distinct intervals

L

load following Operational mode of a power plant that is flexible enough to decrease or increase its output according to demand

load gradient The possible change of a power plant's output over a given time: $\frac{\Delta P}{\Delta t}$, typically given in units of $\frac{\% P_n}{min}$ (percent nominal power per minute)

low voltage Typically 230 V to 400 V

N

n-1 criterion A system that operates with n objects is still functional after the complete loss of 1 object fulfills the n-1 criterion.

network byte order This describes the order in which bits and bytes are transmitted over a computer network; it is the same as *big endian*.

O

Optimal Brain Damage This algorithm modifies the number of connections in an Artificial Neural Network, removing enough that the training set still passes, but the network cannot suffer from overfitting. *See* overfitting.

overfitting This describes that state of a model, specifically an Artificial Neural Network, in which the model describes statistical noise instead of the underlying model and looses its ability to generalize, i.e., to derive meaningful output from a pattern that is not part of the training set. *See* Optimal Brain Damage.

overlay network Is a network existing on the basis of another network architecture as a logic entity on existing infrastructure. Overlay networks often use their own addressing and routing schemes.

P

pari passu Pari passu is a Latin phrase that means 'on equal footing' and can be translated in the sense of 'ranking equally.' When two actors, such as agents, are pari passu, neither actor controls the other one and both have similar tasks of equal importance.

peak load A period of simultaneous, strong consumer demand for electric power

peer-to-peer network An overlay network in which all peers are equivalent; distinct roles such as client, server, and especially router do not exist in peer-to-peer networks as the interconnected peers perform these roles equally

perceptron Denotes a form of Artificial Neural Networks in which input is strictly feed-forward propagated from the input to the output layer, forming a (simple) associative memory.

ping Ping is the name of a network administration software utility program that is used to test whether a host is reachable or not. The name is inspired by the sound of a sonar.

power to gas Usage of electric power to generate and store a form of gas that can be used to drive a power plant

PQ bus (power system load flow analysis) A load bus at which active power and reactive power values are known

prosumer A portmanteau of 'consumer' and 'producer' that describes an entity that can act both, as a consumer and a producer, at different times

pseudo-Boolean function A function that maps Boolean arguments to integer results

PV bus (power system load flow analysis) A generator bus, supplying active power and voltage

R

reactive power Reactive power is power that, due to the delay between voltage and current, cannot perform any work at the load and flows back to the generator.

S

secondary customer Consumers receiving electric power at sub-kilovolt voltages; mostly private households

self-healing A system that is self-healing can automatically reconfigure itself in order to mitigate the effects of a fault.

sensor (agent) This denotes any device, hard- and software alike, that provides the agent with data about its environment in order to allow it to update its internal state.

separation of concerns This is a design principle in computer science that mandates separating a piece of software or a software architecture in several distinct units, where each unit addresses exactly one concern. A concern is a set of information that affects the behavior and, therefore, the code of a computer program.

slack bus *See* VD bus.

smart grid This term denotes a form of the power grid that makes use of digital information, digital information processing, and/or digital communication in order to improve efficiency, reliability, and security of the electric grid.

smart meter A smart meter is a monitoring device that records the flow of power to and from a customer at intervals of an hour or less and transmits the data to the utility for billing. Smart meters can be used to control (e.g., switch on or off) other devices at the customer's site for demand response.

steam generator The boiler of a power plant along with the pipes, valves, etc. that generates steam for the turbines

T

tick Abstract time unit in a discrete-event situation, marked by the occurrence of one or more events

tile server This is a type of server that renders a map at different scales according to a stylesheet and serves the resulting images subdivided into tiles to a client program.

transmission grid This denotes the highest-voltage part of the grid that serves to transmit power over wide distance. Voltage is high so that the current can be low and losses are within acceptable limits.

trusted third party In cryptography, two parties often need to authenticate each other. They do so by each checking the respective certificate of their partner. If the certificate is signed by an issuer whom both trust, the authentication succeeds. Trust is therefore established through a third party that both trust, the trusted third party.

turbine set A number of turbines (high, medium, and low pressure) that are driven by steam coming from the steam generator and excite a generator to produce electrical power in a power plant's block

U

unit test A software testing methodology in which individual units of the software are tested separately, without interaction between each other

V

variable-speed (wind turbine) This term describes a wind turbine design
in which the generator is decoupled from the power grid and can therefore
be driven at variable speed.

VD bus (power system load flow analysis) A bus at which voltage magnitude and voltage phase angle are known

Typographic Conventions

This work will follow certain typographic conventions in order to distinguish termini technici, URIs, source code, etc. from running text.

Technical terms will be printed in italics when first mentioned, but not formatted differently from the surrounding text upon later occurrences. Names of algorithms will be formatted in small capitals.

Whenever a term has an abbreviation, it is mentioned in parentheses upon the first occurrence of the term. After that, only the abbreviation will be used.

Double quotation marks are used to denote quotes from external sources. Single quotation marks, in contrast, denote commonly used terms that are not termini technici or to express statements coming from the author.

Source code will always be printed in non-proportional (typewriter) script.

Additionally, symbols in formulae follow a certain formatting to distinguish constants, real numbers, complex numbers, vectors, sets, and matrices from each other. These are listed in Table 3. The only notable exception from the general rules that are established in this table are the traditional symbols for physical units, such as P for real power, which are not sets, but nevertheless typeset in uppercase italics.

Additionally, Table 3 on the following page shows all typefaces used throughout this thesis to denote words with special meanings.

Table 3: Typographic conventions

Typeface	Meaning	Example
Single Quotes	Commonly used terms	'if and only if'
Double Quotes	Quotes	"To be or not to be [...]"
Italics	Terminus Technicus	Wind energy not used for generating electricity is known as *discard work*.
Typewriter script	Source Code	`puts "Hello, World!"`
	Uniform Resource Identifier (URI)	`http:// www.google.com/`
Upright (in formulae; including parentheses)	Functions	$\mathrm{d}(p, q)$
Upright (in formulae; no parentheses)	Contants	k
Bold italics (in forumlae)	Vectors	$|\boldsymbol{q}|$
Bold (in formulae)	Matrices	$\mathbf{I} = \mathbf{Y} \cdot \mathbf{V}$
Uppercase italics (in formulae)	Sets	$s \in S$
Underlined (in formulae)	Complex numbers	$\underline{Y} = \underline{Z}^{-1}$

1   Introduction

1.1   Motivation

'Somewhere, there's always wind blowing or sun shining.' This maxim could lead the global shift from fossil to renewable energy sources, suggesting that there is enough energy available to be turned into electricity. And there are impressive numbers available: As of today, a number of countries satisfy more than 50 % of their energy demand with renewable energy sources. Iceland, for example, draws all its electricity from them; other countries also have high percentages, such as Norway (92 %) or Brazil (82.7 %) (Observ'ER, 2013).

Closer inspection of statistic reports, however, reveal that these high numbers are backed by generation from hydroelectric power plants, with biomass energy far behind. Neither option, however, is applicable to all countries. Hydroelectric power plants obviously need their construction requirements fulfilled so that the turbines can be driven by strong currents. Biomass is often considered to lead to monoculture in agriculture where plants are grown exclusively for conversion into electricity at the expense of food-yielding crops.

Still, the *European Union* (EU) targets a continuous increase of electricity generation from renewable energy sources to a total of 20 % by 2020 (European Parliament, Council, 2009); the *European Renewable Energy Council* (EREC) has even published a whitepaper that targets a complete supply of electricity from renewable energy sources by 2050 (Zervos et al., 2010). These goals are even more important considering the fast-closing action window for an effective reduction of greenhouse gases (FAZ.NET, 2015).

Where hydroelectric or biomass are not available as sources of electricity, two other, well-known types of technology are brought into focus, which the leading statement implicitly mentioned: wind turbines and photovoltaic panels. Germany, for example, drew 27.8 % of its demand in 2014 from wind power and solar radiation (AGEB, 2015). Although quite easy to harvest, wind and solar radiation make us depend on a phenomenon we cannot yet control: the weather.

Historically, the power grid had to accommodate for a variable power consumption, while the operator had control over the generation of electricity. The grid's capacity, contracts with the owners of power plants, and technical characteristics of the power plants themselves limited the operator's ability to react to changes in the grid. Approximate forecasting through synthetic load profiles allowed for a certain amount of planning ahead of time. These synthetic profiles are created for the year ahead and thus allow for the formation of contracts and general operating plans with a certain time buffer.

The feeding of renewable energy sources has also been included into these synthetic profiles. However, today, no method exists that would allow a precise forecasting of wind and cloud cover for a year ahead.

Inaccurate forecasts can obviously lead to two kinds of error: Either an under-estimation of the feeding, or an over-estimation of the generated electricity. Both can endanger the grid, either by overloading it or by causing outages.

Reacting to over-production is, of course, quite easy; one can simply disconnect wind turbines, whole wind farms, and solar panels likewise from the grid. Although it saves the latter, it also means that potential electricity is lost. This so-called *discard work* is an inefficiency whose primary effect is a financial impact on the owners of those wind or solar farms. In the end, somebody has to pay for this loss of profit. Governments can and do attenuate this impact. However, a loss is a loss; it does not vanish if it is spread out to the taxpayer (Bundesnetzagentur für Elektrizität, Gas, Telekommunikation, Post und Eisenbahnen, 2014).

One could try to store the generated electricity. But no matter what technology is used, be it batteries, pump storage or pressurized air, a non-dismissable percentage of the original energy is wasted in the process of converting it multiple times and large power storage facilities with sufficient capacities and acceptable efficiency are large, hard to build, expensive, or a combination thereof, and have therefore not yet found deployment in capacities where they could supply a large portion of consumers for more than a couple of minutes.

This question of inefficiency due to the inability to put the potential electricity to use poses a question: Could somebody have used this additional energy? Or, better yet: Could somebody have used it, had he known beforehand that it was to become available for a certain period?

Integrating the consumer into the planning process seems to be a logical choice. If the source of energy is not completely under our control, we need to become more flexible in the way we use power. This becomes especially important in cases of a power shortage. Then, it becomes not just a question of efficiency, but one of preventing a collapse due to brown- or blackouts.

Smart meters have been introduced as a means of achieving a shift of load when the generation capacity comes close to exhaustion. The concept is quite simple: What if customers could delay their consumption by a small amount of time, e.g., an hour, in order to take some pressure from the grid? Air conditioning units, on a hot day, could be flexibly controlled to let the temperature rise by 2 °C in order to relieve some pressure off the grid. This way, the loss of comfort for an individual consumer would be minimal, but the sum of all minor power savings would help to soften the spike in the load curve.

The appeal to the consumer stems from a more direct pricing, tailored to the actual situation on the grid. The smart meter will receive continuous updates on the current electricity prices. Together with user-set thresholds, the device can exert control on connected devices in the household.

Fox-Penner (2010) describes this approach detailed in his book "Smart Power." He summarizes an experiment in Sequin, Washington, in 2005, where smart meters had been deployed in a number of households to test this theory: That direct, real-time pricing allows the peak load to shift in order to smooth the load curve. His overall assessment of a smart grid based initially on smart meters is positive; he continues to outline necessary changes and challenges for utilities in the view of market behavior, legislation, and hardware maintenance.

Reality elsewhere, however, has seen mixed results to this approach. Some field studies suggest that users are inclined to make use of this kind of control in order to save money, while other surveys have received negative feedback, indicating that most people lose interest in this technology after a short period of time (Merrion, 2011; Buchholz et al., 2012). Also, one could argue that electricity should be a 'when you need it'-type of resource.

Making consumers use more energy instead of less is even harder to achieve. That would mean that the smart meter turned on the washing machine whenever electricity is inexpensive. Typically, however, people do not want be forced to do their laundry just because the electricity is cheap, especially if this comes without warning.

Currently, variances in photovoltaic and wind farm production are handled by base load power plants that can be controlled in a timely manner. Wind farms and photovoltaic installations just feed into the power grid; control impulses are scarce.

However, with an increasing use of renewable energy sources, this does not remain an option. Intellegient grid management is necessary in order to accomodate the volatility of these resources. Microgrids with fine-grained forecasts and potential island mode are needed to accomodate the distributed nature of these sources.

But the increase of distributed power generation also increases the amount of information necessary for monitoring and controlling the power grid. Today, central points of control are in charge to ensure the stability of the power grid in general. However, considering the increased load, how do we prevent these single points of control from becoming single points of failure?

1.2   Contribution and Constraints

If an increase in information exchange can lead to more opportunities to include renewable energy sources in a more efficient way, the amount of planning for the short term will also increase. This, however, will in turn also require an efficient handling of this information.

To this end, this work proposes to accompany a more and more distributed power generation system with distributed software that enables consumers and producers alike to engage themselves in a grid-wide planning phase. The idea of handling a large task by breaking it down into smaller ones and distributing the workload over a number of participants is generally not new; the *Contract Net Protocol* Smith (1980) proposes, formulates a high-level approach on a protocol basis for exactly this task. Smith assumes that a task can be subdivided into smaller ones, and that any node can design whether it will take on the large task at hand, or award this task—or parts of it—to other, remote, nodes. These, in turn, can decide whether to claim the task or not, offer a certain price for it, and, after winning the bid, even subcontract it to other nodes. While appealing, this approach needs refining in order to find an application in a smart grid; balancing the grid's power level cannot be optional; a different set of rules applies when contracting for power supply or consumption.

Through localized forecasting that is able to include the characteristics of a certain site, nodes in the grid are enabled to act in a proactive manner. They become *agents*, equipped with one piece of software instanciated on each node that initiates and performs a grid-wide demand-supply calculation. This calculation starts in the immediate vicinity of the acting node, but spreads if the volatility cannot be handled locally.

Therefore, this work views the problem of including renewable energy sources to provide base load functionality as a problem of information interchange and distributed intelligence, i.e., as a software and networking problem. It makes use of the increased efficiency and network connectivity of factories, or the primary control capabilities exhibited by newer installations of wind turbines or Photovoltaic panels. However, it does not promote nor actively propose further development in this area; models of these nodes merely form the basic building blocks for the software agent proposed in this thesis.

The forecasting module of the agent that enables it to act pro-actively will not be based on meteorological models, but on machine learning. Detecting patterns in weather conditions is done using Artificial Neural Networks, to which this work contributes an analysis of a new learning algorithm.

Distributed software needs a means to communicate. A network protocol for this purpose is therefore proposed in this work. It makes use of standard technologies available in the traditional ISO/OSI network stack in order to provide a set of behavioral rules all agents must follow. Information encoding, however, is not part of the proposal, since we can assume that techniques providing efficient information encoding are widely known and thus this dissertation will not contribute to research in this area.

The proposed software agent will be easy to set up. The software must not need extensive, on-site maintenance, since wind farms are often located in remote areas, and, considering off-shore installations, can be hard to reach. The software is therefore self-aware, a trait that also contributes to the justification of the agent title; no modifications are necessary to accommodate different types of generators, transformers, or consumers, making the agent universal with regards to the type of the node in the power grid.

Network communications and self-aware peer-to-peer networks need extensive security and the concept of trust. This thesis will make use of known techniques in order to attain a secure communication, but will not go beyond that. The creation of new security concepts is out of the scope of this work.

The heart of this piece of software is the control algorithm that enables it to act and decide based on requirements received from other agents or stemming from its own power generation or consumption. This can be viewed as a problem purely rooted in the Boolean domain and as such, will be solved with the tools and calculus available in this field.

1.3 Overview

In order to describe the transition from a purely centralized grid control to a distributed, self-organizing grid, augmented by centralized control, the first sections of Chapter 2 will outline the state of today's power grid. This chapter will continue to present the fundamentals required to understand the contributions of the following parts of this thesis; it therefore also outlines essential knowledge considering Boolean algebra, communication networks, and artificial intelligence. It discusses the relevant literature in order to allow the reader to delve more deeply into the corresponding topics if desired.

Chapter 3 introduces the models used by the Universal Agent. The whole design of the agents needs, of course, extensive testing, which can be accomplished by the means of simulation. A simulation of the grid or even only a part of it depends on various input data sources and, in case of renewable energy sources, especially weather data. Measurements differ in their quality; an assessment of data quality is therefore essential. Chapter 3 therefore also describes a simulation environment for which the continuous monitoring of the quality of all input data is an essential feature.

Stringently, the design of the proposed software agent is introduced afterwards. Chapter 4 outlines the modules in use and presents central concepts and assumptions of the design.

In order to act proactively, effective forecasts must be made. Since the agents and their planning is of a distributed nature, the forecasts themselves will also heavily rely on localized forecasts. Each node therefore employs machine learning in order to forecast weather conditions, power generation, or power consumption. The corresponding agent module is the subject of Chapter 5.

Afterwards, Chapter 6 details the communication phase of a demand-supply calculation. Agents are now set up and can use the communication infrastructure; the software now needs binding rules for communication demand and supply. During this phase of the distributed planning, all other modules of the agent are considered to be black boxes. Chapter 6 represents the outside view on the agent.

With a working communication, this thesis turns its attention back to the internals of the agent. The algorithm used to find a solution in the demand-supply calculation is specified in Chapter 7. It is the central of the three pillars upon which the functioning of the distributed software rests.

Finally, Chapter 8 concludes and attests whether the elementary theory of this thesis can be positively asserted: Namely, whether a distributed, proactively acting agent system based on a piece of software that is instanciated on every node is able to achieve a more effective integration of renewable energy sources with high production volatility.

2 Fundamentals and Related Work

2.1 The Electric Power Grid

Power Generation

Fundamentals of Power Generation

The traditional electric grid has, from the bird's eye view, always been understood as having a unidirectional architecture: Power is generated in large power plants and is distributed via transmission lines to customers, where it is consumed.[1] Power generation has always been centralized: with a small number of generating stations supplying a large number of consumers with power. The consumer, in turn, was strictly that: A form of load that needed to be borne. Fig. 2.1[2] depicts this traditional, one-way flow of power.

Power generation mostly relies on electromagnetic induction;[3] the constant changing of a magnetic field creates a current in a conductor. Large-scale power generation utilizes mechanical motion in order to provide a magnetic field changing in a constant and reliable manner. The rotating magnetic core, called *rotor*, induces current in the *stator*, a fixed wire. The stator ultimately interfaces with the power transmission system and the lines that lead from the power plant. Since it is the rotation of the magnetic core that induces the current in the *field poles*, the current alternates with the rotation. The

---

[1]The terms 'generation' and 'consumption' are, in a strict physical sense, wrong. Energy is not generated or consumed, but always converted from one form to another. However, these expressions have been widely accepted; also, the conversion of electric energy to, for example, heat can be thought of as a consumption of electric power in order to convert the energy to another form. Thus, we will also employ the notion of a producer or generator and a consumer.

[2]Based on US-Canada Power System Outage Task Force (2004)

[3]With the exception of Photovoltaic panels, which we will cover later in this chapter.

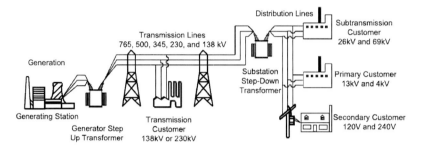

Figure 2.1: The traditional view of the power grid as one-way flow of power from generators to consumers

frequency of this *Alternating Current* (AC) is fixed for the whole power grid; in most countries, like in Europe, at 50 Hz or at 60 Hz, as in the USA or Canada. The frequency of the AC is one of the most important indicators for power quality: Deviating from the set frequency damages devices attached to the grid.

The mechanical motion of the magnetic core is most commonly driven by a turbine. One way to excite the turbine is steam or water, but using gas is also possible, as well as relying on wind energy. The following paragraphs will provide a survey of the most commonly used types of power generating stations. Many books provide a detailed account of the mechanics involved, for example, Heuck et al. (2010), Allelein et al. (2010), and Oeding and Oswald (2011).

For the grid operator, power plants typically fall into one of three categories: *Base load*, *load following*, or *peak load* power plants.

A base load power plant is constantly operated at its maximum efficiency, typically producing its rated power output. A base load power plant is not throttled, and only taken offline due to maintenance. The reason for this type of operation either lies in its design, e.g., it is simply not possible to change the plant's output by much, or that throttling would damage the plant. It might also be harder to start, or take a long time to synchronize with the grid. Additionally, the cost factor of a power plant plays an important role in its categorization: Base load power plants are typically most economically run at their rated power, but quickly become unprofitable if not run in this way. Nuclear power plants are an example of base load power plants.

Load following power plants can be throttled and operated at non-rated power output levels. These power plants can react to changes in the power grid fast enough and with sufficient amounts to take on that role. However, due to their design or for other, mostly economic reasons, these power plants are not disconnected from the grid. This can be due to their design that requires a certain base load in order not to damage the plant, such as coal, and oil power plants. Additionally, these power plants require electricity to start, i.e., they are not *black start* capable, which is another reason why these are considered load following, but not peak load power plants.

Peak load power plants are flexible enough that they can be shut down if not needed as well as started and synchronized with the power grid very quickly in order to react to peak loads. Gas turbine power plants are usually considered peak load power plants.

Hydroelectric Power Plants

Water is used directly in *hydroelectric power plants*. Here, gravity is used through a downward slope to create a flow of water that applies kinetic energy to the turbine. Hydroelectric power plants are quite simple in their overall design as the turbine is directly driven by water and coupled to the generator. The turbine design changes with the head of water. For heights of 60 m or less, a hydroelectric power plant, then called a *low-pressure unit*, is constructed with a *Kaplan turbine*, which resembles a propeller. Kaplan turbines provide a uniform flow speed over the whole surface of the turbine, which is important to keep the mechanical stress to a minimum. Its impeller vanes are also adjustable, allowing it to adapt to the water flow. Hydroelectric power plants with heads of water between 60 m and 300 m, called *medium-pressure units*, feature a *Francis turbine*. Here, the water flows through a circle of guide vanes and hits the turbine radial instead of directly from above. *High-pressure units*, defined by a head of water of more than 300 m, usually requires a *Pelton turbine*, where water flows through cones to the turbine. Synchronous machines ensure that the output frequency of the power plant is stable at the grid frequency.

Hydroelectric power plants can exist as run-of-river power stations, where a barrage in a river is used to drive Kaplan turbines. Storage power stations as depicted in Fig. 2.2 use a natural reservoir and have a higher head of water. In order to keep pressure stress low on the tubes leading to the turbines, when the vents are closed quickly, so called *surge tanks* at the reservoir provide the required pressure compensation. Storage power stations can be constructed to fill their reservoir with pumps when excess electric power is available in the grid, allowing to store it for a later time when more power is required. The efficiency of such a pump storage power plant currently peaks at $\eta = 83\%$ (Sterner and Stadler, 2014, p. 492).

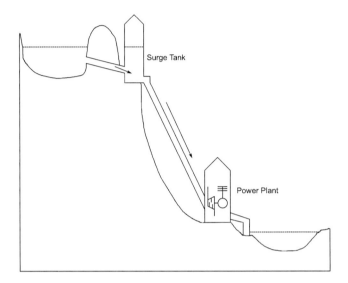

Figure 2.2: Schema of a storage power station

This type of power plant has two important features regarding its role in the power grid. First, it is easy to start and bring online. In order to put it into service, only the gate valve needs to be opened. The pressure of the water streaming in drives the turbine, thus generating power. Hence, a hydroelectric power plant does not depend on an external source of power in order to start. This feature is called black start capability.

The second important feature of a hydroelectric power plant also stems from its relatively simple design: It can change its power output faster than condensation power stations such as coal- or oil-fired plants. It can be brought online from complete standstill in one to two minutes (Heuck et al., 2010, p. 24); it can change its output by 100 % to 200 % of its rated (nominal) power per minute (VDMA Power Systems, 2013, p. 16). This second characteristic is called the power plant's *load gradient*[4] and given in units of percent nominal power per minute: $\%P_N/\min$.

For a power grid operator, it is these characteristics that define a power plant: The time it needs to start and to be brought online, its ability to start without power from the grid—i.e., whether it has black start capability or not—,

---

[4]I.e., $\frac{\Delta P}{\Delta t}$

its nominal and, possibly, minimal power output, and its load gradient. When we now discuss other types of power plants, we will see that these values vary by a great degree based on the technology used to generate power.

Steam-based Power Plants

A large portion of power plants use a thermal energy derived from another, primary energy such as chemical energy from coal, or oil, or nuclear fission. Water is heated; the steam drives the turbine. The steam is then cooled, again fed to the boiler, where it is again turned into steam. This form of power generation ensures a stable, continuous supply of electricity; the primary sources of energy—coal, oil, uranium, etc.—provide independence from volatile external factors such as weather conditions.

To describe these power plants as 'glorified boilers' is, however, an over-simplification that disregards the characteristics of the design of each power plant type. Power plants are not only categorized by the primary energy resource they rely on, but also by the behavior and the amount of control an operator can exert on the power plant based on the primary energy it uses.

We will now cover power plants that utilize fossil primary energy sources, namely coal, oil, and uranium, before outlining the functioning of gas turbine power plants.

Any turbine-based power plant is subdivided into a number of *blocks*: Each *steam generator* or boiler drives a *turbine set* that excites the generator. The maximum amount a generator can output is called its rated output and serves to further characterize a block. For example, a '1000 MW block' is a unit of a steam generator, turbine set, and a generator that can continuously feed 1000 MW into the power grid.

Coal power plants use the chemical energy bound in coal to generate thermal that is, in turn, converted into mechanical energy. A schematic view of such a coal power plant, based on Heuck et al. (2010, pp. 19 and 21), is presented in Fig. 2.3 and accompanies the following description.

The burner of a block converts the chemical energy of the primary resource to thermal. In a coal power plant, a mill turns the coal into coal-dust, which is blown into the furnace of the burner. Hot air is blown in from below to feed the fire.[5] The fire heats water that flows through pipe runs. The feed water pumps create high pressure between 170 bar to 330 bar (Oeding and Oswald, 2011, Fig. 3.2d). The steam drives a number of turbines, divided into high-pressure,

---

[5]This air is heated by the exhaust of the furnace.

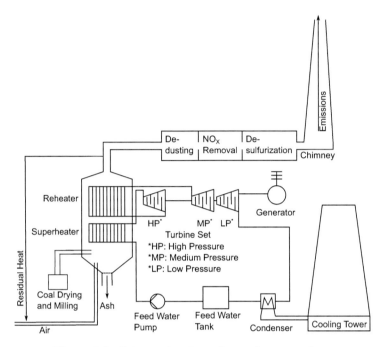

Figure 2.3: Schematic view of a coal power plant

medium-pressure, and low-pressure parts. Finally, a condenser cools the steam back down to water, with the help of a nearby river and a cooling tower. On its way back to the feeder pump, the water in this closed loop is pre-heated using bled steam feedwater heating, i.e. using steam that has been tapped after the first turbine.

The higher the values for pressure and temperature are during vaporization, the higher is the block's efficiency. Obviously, the quality and durability of construction materials forms the direct limit. Raising the values of the main steam's state variables from $p = 160\,\text{bar}, \vartheta = 530\,°\text{C}$ up to $p = 280\,\text{bar}, \vartheta = 600\,°\text{C}$ increased the block efficiency from $\eta \approx 38\,\%$ to $\eta \approx 47\,\%$ (Heuck et al., 2010, p. 7).

The construction materials, mainly of the pipes that carry the steam as well as those of the boiler, also determine a steam power plant's time until it can be brought online, and also limits its load gradient. During the start-up phase, the boiler grows up to 30 cm (Heuck et al., 2010, p. 10). That is why grid operators estimate a coal power plant's start-up time according to the time it has been offline: Less than 8 hours, 8–48 hours, and more than 48 hours.[6] A coal power plant uses about 5 % of its output for its own requirements, such as the coal mills or the pumps. Therefore, it has no black start capability.

The operator controls the plant's power output with the drive of its turbines, which stems from the amount of steam that is directed to the turbine's vanes and that is increased or decreased by valves. The fireing of the boiler must match the desired flow rate of steam. The transfer of heat from the firing to the water is the plant's limiting factor regarding its load gradient, not only in terms of speed, but also because of potential damage to the pipes. The latter is the reason that a steam power plant also must run with a certain minimal load: The pipes must be evenly heated in order to avoid fissures due to heat stress. This minimum load depends on the plant's construction details and varies between 35 % and 65 % (VDMA Power Systems, 2013, p. 16). Another factor that limits the load gradient is the fuel itself: Anthracite and bituminous coal that is extracted in mining operations have a higher heating value[7] than sub-bituminous coal and lignite,[8] the latter being won in opencast pits. For an explanation on the different sources of coal, its formation, and properties, refer to Ghosh and Prelas (2009, Chapter 6).

A steam power plant can also generate heat using oil, which changes the configuration of its burners, but not the general principle of power generation. From that point of view, a nuclear power plant also does not differ much, since the way it generates electrical power is also by means of steam driving a turbine. However, the design considerations vary greatly due to the different nature of its fuel: Nuclear fission is a process that is not easily interruptible. It also takes more time for a nuclear power plant to come online after complete standstill. Additionally, a nuclear power plant requires electricity even after it has been shut down: The fission process cannot be abruptly stopped; heat is generated even after shut down. This heat, called *after-heat* or *decay heat*, requires the reactor to be cooled after it has stopped generating power, thus causing the power plant to require power itself.

There are many different designs for nuclear power plants available. They focus on cost optimization, increase of efficiency, rated power output, or safety.[9] They are usually classified by the moderator they use, i.e., by the compound, and application thereof, that controls the nuclear fission, and by the coolant

---

[6]See, for example, VDMA Power Systems (2013).

[7]Between approximately 32 000 kJ/kg and 36 000 kJ/kg

[8]Approximately 28 000 kJ/kg

[9]Not all these parameters are mutually exclusive.

and cooling circuit design they employ. For the power grid operator, the same variables are relevant as with other steam-based power plants: The start-up time depends on the plant's offline time, there is a minimum load required, and the load gradient resembles that of other steam-based power plants. For a detailed outline of different nuclear power plant designs, refer to Ghosh and Prelas (2009, Chapters 9.10 to 9.16).

These steam-based power plants have, in contrast to the aforedescribed hydroelectric power plants, traditionally been considered to be 'base-load behemoths.' However, recent development has made authors argue that these power plants should also participate in the power grid's load management in a flexible manner (Brauner et al., 2012), which will change at least the role of coal- and oil-based power plants.

Gas Power Plants

Power plants fueled by natural gas do not drive the turbines using steam. Instead, they work like a jet engine: Fresh air is fed via a compressor ($p = 15..20\,\text{bar}$) to the combustor, where it is used to burn the gas. The exhaust ($\vartheta \leq 1500\,°\text{C}$) drives the turbine directly. The turbine drives a generator as usual. The exhaust is afterwards released to the open, which is why this way of operating a gas turbine is also called *open gas turbine process*. This direct method of driving a turbine in order to generate electrical power allows the plant to start very quickly; ceramic shielding in the combustion chamber, temperature-stable monocrystal impeller vanes, and film-cooling of shields and vanes with compressed air allows a significantly greater load gradient compared to traditional steam-based power plants[10] while being more expensive to operate.

Additionally, power plants of this type are capable of performing a black start, and can be brought online in seven minutes.

The exhaust of the turbine is still relatively hot[11] and can therefore be used to additionally drive a traditional steam turbine. Here, the exhaust powers a superheater that generates steam. This combined gas-and-steam process is shown in Fig. 2.4.[12] Although the overall load gradient of the combined power plant is lower than that of a pure gas turbine power plant due to its steam part, the gas turbine can be operated independently, if necessary.

---

[10] 10 % to 25 %, cf. VDMA Power Systems (2013).
[11] $\vartheta \leq 625\,°\text{C}$, cf. Heuck et al. (2010, p. 21).
[12] Cf. Heuck et al. (2010, pp. 19 and 21).

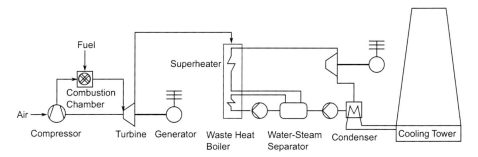

Figure 2.4: Schematic view of a combined gas and steam power plant

Geothermal Power Plants

A renewable and, in terms of emissions, 'clean' energy source is geothermal
energy. The power plant extracts some of the Earth's heat content via drill
holes to power a steam turbine. Different layouts of power plants exist, based
on the heat of the drillhole. The *Intergovernmental Panel on Climate Change*
(IPCC) Special Report on Renewable Energy Sources and Climate Change
Mitigation (Goldstein et al., 2011) outlines the types commonly used today.

A *condensing steam* power plant is constructed for intermediate or high-
temperature sites ($\vartheta \geq 150\,°C$). The pipeline from the heat source leads to a
steam-water separator; the steam drives a turbine set (turbo generator); the
turbine set is connected to a condenser and cooling tower. From the steam-water
separator, a second pipeline leads back to the reservoir via an injector well,
pumping the separated water back in order to obtain more steam, thus forming
a cycle. When the well delivers hot water, it 'flashes' to steam; in contrast, a
dry well delivers steam only.

Lower temperatures[13] in the well are accommodated by *binary cycle* units.
They are more complex than the condensing power plants since the geothermal
fluid is led to a heat exchanger where it heats a second working fluid with a
low boiling point[14] that drives the turbine—hence the name 'binary cycle.'

Geothermal energy is considered a sustainable form of energy, as long as
the amount of fluid extracted from the well stays within the limits of what the
source can produce (Rybach, 2007). The ambient temperature does influence
a geothermal power plant (Imroz Sohel et al., 2009) and can be a target of

---

[13]Low- to intermediate-temperature fluids, i.e., with temperatures from $70\,°C$ to $170\,ř C$

[14]E.g., isopentane or isobutene

optimization, but is normally considered negligible (Rybach, 2007). The cost of finding a suitable site, drilling holes, and generally erecting the power plant are relatively high[15] compared with other power plant types. Therefore, a geothermal power plant is profitable and efficient as a base load power plant. Due to their design, geothermal power plants obviously are black start capable.

Wind Turbines

Another renewable energy source is wind. In this short survey of power generators, it is the first volatile energy source: It relies on the current of the wind to generate power, a force that, especially onshore, is not steady but changes with weather conditions, sometimes rapidly. Wind turbines cannot increase their power output on demand, unless they have previously been throttled in order to allow for a load-following capability (MacDowell et al., 2015). If no wind is blowing, it obviously cannot produce any power at all. Thus, the output of a wind turbine is inherently fluctuating; it cannot be dispatched.

Historically, with windmills, this is one of the oldest forms of energy mankind has employed as an aide to his work. Over time, different designs of wind turbines have emerged; however, the *Horizontal Axis Wind Turbine* (HAWT)[16] with three blades is the most commonly seen today. This is due to the reliability of the three-bladed design, their cost-effectiveness compared to other designs, and finally their efficiency. This survey will continue in its operator-centric view; the interested reader should refer to Manwell et al. (2010), which is a comprehensive monograph covering the historical perspective, different wind turbine designs, the physics of wind force usage and material design, as well as economic and environmental aspects.

Most wind turbines are two- or three-bladed, which is due to the fact that, in order to achieve the greatest efficiency, a wind turbine needs a design that does not cause the wind to flow around it like an obstacle, but to pass through it in order to apply its force to the rotor. The ideal wind turbine and its maximum efficiency[17] has been modeled and proven in 1926 by Albert Betz (Betz, 1926). The three-bladed design provides more stability and runs more steadily than the version with two blades. Almost all three-bladed wind turbines have upwind configuration, i.e., the wind reaches the rotor first and then the tower.

The rotor and the configuration of the wings contribute most to the controllability of a wind turbine, and are "often considered to be the turbine's most

---

[15]Cf. also Goldstein et al. (2011).

[16]Meaning that the axis of rotation is horizontal, i.e., parallel to the ground

[17]The *Betz limit* of $\eta = \frac{16}{27} \approx 59.3\,\%$

important components from both a performance and overall cost standpoint" (Manwell et al., 2010, p. 4). The goal of turbine control is to maximize energy production,[18] preventing extreme loads as to minimize fatigue damage, to provide acceptable power quality, and to ensure safety while operating the turbine (Manwell et al., 2010, p. 370). This means that wind turbines are, whenever possible, operated at their rated power.

How the rotor and its blades are used to control the wind turbine's overall power output depends on the mechanical-technical configuration of the drive train, gearbox, and generator in the nacelle. A distinction is made between *constant-speed* and *variable-speed* operation.

On constant-speed wind turbines, the generator is directly connected to the grid, i.e., the power grid's frequency dictates the generator speed, similar to the connection of steam-based power plants. Stringently, this dictates the rotation speed of the blades too, to which the term 'constant-speed turbine' refers.

Accordingly, a variable-speed wind turbine allows for varying rotational speeds. One design approach of variable-speed wind turbines uses a synchronous generator driven by the rotor. In order to decouple the frequency of the turbine from that of the grid, the generator is connected to a rectifier that converts the alternating current of variable frequency to *Direct Current* (DC). The rectifier feeds the DC to an inverter, where it is converted again to AC with the frequency of the power grid. Two other, conceptionally less straight-forward designs use squirrel cage induction generators, or wound rotor induction generators, respectively. While the former tries to increase the variable-speed turbine's efficiency, the latter facilitates variable-speed operation by offering so-called *true variability*.[19] For a more detailed discussion of the design approaches, refer to Manwell et al. (2010, Chapter 5.6).

Variable-speed wind turbines keep the stress to the material at a minimum while generating power efficiently at different wind speeds. In theory, however, fixed-speed wind turbines are more efficient, buying this advantage with increased stress on the rotor and drive train as well as the need to design rotor and blades with mechanisms to change their geometry in order to adapt to different wind speeds while still holding the rotational speed of the turbine constant or near-constant. While Carlin et al. (2003) stated that most wind turbines are constant-speed turbines, the simpler blade design of variable-speed turbines compared to the need for variable-geometry blades of constant-speed turbines provides an advantage in terms of material stress and improvements in

---

[18]below rated wind speed

[19]Here, power is fed into the rotor, allowing it to work at sub-synchronous speeds.

power electronics have made grid connected variable-speed turbines a preferred alternative of many recent, larger wind turbine designs.[20]

In order to exert control over the turbine, several modes of operation exist. Constant-speed turbines are either stall- or active pitch-regulated. The blade design of a stall-regulated wind turbine is such that it passively regulates the power production of the turbine. The blades are fixed-pitch; the wind's angle of attack increases with increasing wind speed. Because of that, a growing part of the blade, starting at the blade root, enters the stall region. This effectively limits the wind turbine's power output and requires heavier blade structures[21] that can withstand the blade-bending loads that are typical of this design (Manwell et al., 2010, p. 372). Active-stall designs use the same physical effect, but here, the blade can be rotated at the hub about the blade axis. The blades turn to the front and out of the wind.

Active-pitch regulation also works by turning the blade about its axis, but here, it is turned downwind. This method of regulation features a higher angle of rotation and works faster.

Variable-speed wind turbines use active-pitch regulation; stall-regulated variable-speed turbines are a topic of research, but no commercially viable designs have emerged yet.[22] Active-pitch-regulated wind turbines operate with fixed pitch in part-load situations in order to optimize the tip speed ratio.[23] When rated power is reached, pitch regulation is employed in order to keep rotor speed within an acceptable limit. The generator torque is used to control the power output. This configuration maintains a constant power output during gusts while the rotor speed increases. The energy of the wind is stored as kinetic energy in the rotor. Diminishing wind speed leads to reduced aerodynamic torque while the generator output is held constant; if the wind speed remains high, however, the pitch mechanism reduces the aerodynamic efficiency and with it the aerodynamic torque.

To a grid operator, the cut-in and cut-off wind speeds are important: They describe when the wind turbine is brought online or when its control mechanisms

---

[20]Cf. Vestas' 2 MW wind turbine products (Vestas Wind Systems A/S, 2012), but also Enercon's smaller products (ENERCON GmbH, 2011).

[21]Either heavily welded, or cast structures

[22]Cf. Manwell et al. (2010, Chapter 8.3.2.1).

[23]The *tip speed ratio* denominates the ratio of the blade's tip speed to the wind speed:

$$\lambda = \frac{\omega R}{v} \ ,$$

with ω being the rotor's rotational speed in radians per second, R being rotor's radius in meters, and v denoting the wind speed in meters per second.

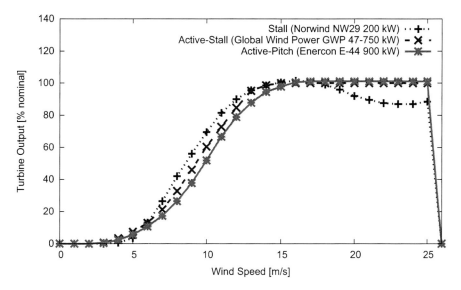

Figure 2.5: Power curves of wind turbines with different control technologies

stop the turbine in order to avoid damage. These wind speeds can be plotted on a power curve that maps wind speed to a wind turbine's output. Such a plot also shows the effect of the different regulation designs. Three different wind turbines of different manufacturers with different regulation mechanisms are shown in Fig. 2.5. Both cut-in and cut-off wind speeds are typically averaged over a number of minutes, i.e., a wind speed of 5 m/s that lasts for 1 minute does not cause the wind turbine to synchronize its generator with the power grid. Obviously, this is aimed at reducing fluctuations in the power grid from wind turbines that are brought online and offline in quick succession. The fact that a wind turbine needs no external power except for wind makes it capable of performing a black start.

Another regulation mechanism of modern wind turbines is concerned with the wind's direction. In order to achieve maximum efficiency, the rotor's surface of revolution and the attack vector of the wind must obviously form a 90° angle to each other. The turbine's whole nacelle is rotatable by 360° in order to achieve this. More specifically, the nacelle can be turned by more than 360°, e.g., two full turns. How many turns a nacelle can make in order to 'follow the wind' is specific to a manufacturer, however, and is limited by the cabling in the tower, which is being twisted by the wind-following motion of the nacelle. After this limit has been reached, the wind turbine needs to shut down and employs a motor to unwind. After that, when it has reach its 0° position, it resynchronizes itself with the grid.

The wind turbine regulation options presented here are important for the turbine to be reliably operable while converting as much wind energy to electrical energy as possible. However, they are not employed to throttle the wind turbine's output. Instead, a whole wind farm, i.e., a collection of wind turbines is subject to the grid operator's control. Reducing the farm's output is achieved by throttling select turbines or by bringing them offline.

It is obvious that, while the regulation mechanisms presented here try to allow a turbine's output to be at its rated power for as long as possible, a larger part of the power curve is highly dependent on the current wind. Forecasts thus play an important role when increasing the share of wind power in the energy mix. We will outline one way of forecasting using local data in Chapter 5.

Asides from its fluctuating active power output, wind farms have traditionally been considered grid-burdening generators, because they require inverters to generate AC and therefore influence the balance of reactive power and the operator's ability to dispatch reactive power when needed. However, modern wind turbines can also offer *ancillary services* (MacDowell et al., 2015).

Photovoltaic

Photovoltaic (PV)[24] power plants use photovoltaic panels to create electrical power from solar radiation. For the layout of these panels, two noteworthy designs exist today.

The older variant uses doped semiconductors of the periodic table's IVth main group that release charge carriers when irradiated with light. This is called the *photoelectric effect* and in this design is achieved by using silicon[25] that is doped with elements of the IIIrd or V[th] main group. This way, n-type and p-type semiconductors are created; layers of n- and p-type semiconductors further create the n-p transition. Here, an electric field is developed that separates the charge carriers of the layers. This creates the DC of the cell. The cells are connected in series in order to achieve the desired voltage.

The creation of crystalline cells is material consuming, since the cells are cut from one block of silicon and feature a thickness of about 400 μm. A better

---

[24]The word is artificially created from the greek work for light, *phos*, genitive *photos*, and the term *voltage*.

[25]The semiconductor

alternative in this context are thin-layered solar cells made from copper, indium, gallium, and (di-) selenide, which are only $3\,\mu\mathrm{m}$ thick. Moreover, the fabrication of silicon-based cells is more energy consuming: A silicon cell typically has to work for 1.5 years to 2.5 years, depending on the amount of solar radiation received every year, in order to produce the amount of energy that was required to create it (Burger et al., 2016).

In order to judge the performance of a solar cell, two values are of importance: first, its efficiency and second, its specific size ratio. The latter is expressed in terms of $\mathrm{m}^2/\mathrm{kW_p}$, where the index p stands for 'peak,' i.e., the peak output is reached at $E = 1000\,\mathrm{W}/\mathrm{m}^2$ radiation power.

In past years, polycristalline silicon cells have been considered to have the highest efficiency. For example, Heuck et al. (2010, Chapter 2.4.8.1) lists mono-crystalline cells with $\eta = 14\,\%$ and a size ratio of $7\,\mathrm{m}^2/\mathrm{kW_p}$ to $9\,\mathrm{m}^2/\mathrm{kW_p}$, multi-crystalline cells have been rated with $\eta = 13\,\%$ and a size ratio of $8\,\mathrm{m}^2/\mathrm{kW_p}$ to $11\,\mathrm{m}^2/\mathrm{kW_p}$, whereas *Copper Indium Gallium (Di-) Selenide* (CIGS) modules have been quoted with $\eta = 10\,\%$ and $11\,\mathrm{m}^2/\mathrm{kW_p}$ to $13\,\mathrm{m}^2/\mathrm{kW_p}$. More recently, Burger et al. (2016) tally monocristalline silicon cells at $\eta = 22.9\%$, polycristalline silicon cells at $\eta = 19.2\%$, and CIGS modules at $\eta = 17.5\%$. Through recent progress in CIGS development, an efficiency of $\eta = 21.7\,\%$ has been reached under laboratory conditions (Jackson et al., 2014), while the best polycristalline silicon cell currently peaks at $\eta = 21.3\,\%$ (Burger et al., 2016).

To the grid operator, the fluctuation of an array of solar modules due to solar radiation is of importance; the peak output is determined by the actual modules that are being used. Thus, forecasts are as important for PV installations as they are for wind farms. A PV power plant is capable of black start as long as the sun is shining.

Other Means of Generating Power

The power generator types presented in the previous sections are not the only ones that are in use world-wide to produce electric power. Those are, however, the most commonly encountered types.

Other types of power plants use tidal force. A reservoir is filled during high tide; upon low tide, the water flows back into the sea, driving a turbine or a set of turbines with it. Those types of power generators must be seen as base-load power plants since they have limited controllability by design. The reservoir must be sized in order to supply water for the whole six hours of low tide. For example, consider a power plant that supplies $P = 1000\,\mathrm{MW}_{el}$ with an efficiency of $\eta = 80\,\%$. If the average slope usable, i.e., the difference between reservoir and the sea, is $\Delta z = -4.5\,\text{m}$, it requires a volumetric flow rate of $Q = 2.8 \times 10^4\,\text{m}^3/\text{s}$. If the reservoir's water level will vary by $1\,\text{m}$, the reservoir will require a surface area of $600\,\text{km}^2$ (Allelein et al., 2010, p. 267). Kaplan turbines with adjustable vanes can be driven with water flow from both directions, i.e., the plant will generate power also during high tide.[26]

Engineers also use the power of waves for generating electric power. Researches have classified waves into three categories: seismic, surf, and wind waves. The latter two's energy can be converted into electric energy by using the continuous alternation between potential and kinetic energy. That can be done using floaters whose periodic up-and-down movement drives a generator, through sub-sea installations where the movement of the waves is used to pump a work fluid that actuates a generator or an Oyster-resembling construction where two frames are connected by a hinge: the lower frame is fixed to the ground, whereas the other one is moved by the waves. The movement drives a pump cylinder, driving air with $p = 69\,\text{bar}$ through a tube to a turbine installed on the land.[27]

In terms of using solar radiation, solar heat power plants are in use. They capture solar radiation using collectors, focusing the rays to heat a work fluid. Considered base load power plants, they are obviously dependent on the sun and capable of performing a black start.

Power Storage

In our simple producer–transmission–consumer view, power storage is not easy to place. One can argue that, since the efficiency of any known method of storing power is below $100\,\%$, a power store is more a consumer than it is a producer. Additionally, it does obviously not 'produce' power by itself, but rather conserves power produced by other generators. Still, the purpose of any form of energy storage is to store power that existed in surplus at a point in time and could not be used by consumers in order to re-feed it into the power grid when needed. Therefore, we will survey forms of energy storage available today in this section, all the more as experts consider energy storage to be a necessary key point with regards to a high penetration of the power grid with renewable, volatile energy sources.

The hydroelectric power plant with a reservoir that is pump-filled when a surplus of power exists, has already been presented in Section 2.1. Other means

---

[26]Cf. Allelein et al. (2010, p. 268).
[27]Cf. Allelein et al. (2010, Chapter 11.7).

of energy storage can be roughly categorized as either electrical, electrochemical, chemical, mechanical, or thermal, based on how power is actually stored. This survey will outline the different types of storage in the order mentioned here, and briefly discuss their properties regarding their inclusion and use in the power grid. It is based on Sterner and Stadler (2014), a—of the time of writing—current and extensive monograph covering current power store technologies.

Electrical energy stores include so-called *supercaps*, i.e., cylindrical capacitors that use a dielectric that is rolled in order to increase the overall area, thus increasing the capacity of the supercap.[28] Supercaps have an energy density of $w = 0.1..10\,\mathrm{Wh/kg}$ at $\eta = 90..95\,\%$, making them useful at places where high wattages are needed for a short period of time, for example, in wind turbines for pitch regulation. Superconducting coils are another way to store electric energy without prior conversion. Even though they, too, possess a high efficiency of $\eta = 92\,\%$, the energy density of superconducting coils is low at $w = 1\,\mathrm{Wh/kg}$. Additionally, the cost of the two energy store systems is high compared to other methods: supercaps cost $5150\,\text{€/kWh}$ to $12\,000\,\text{€/kWh}$, superconducting coils $13\,570\,\text{€/kWh}$ to $75\,670\,\text{€/kWh}$. These properties make them only for niche applications the preferred choice, but not for the overall concept of a power-grid-supporting energy storage.

Electrochemical energy storages denote the colloquially known 'batteries.' Of the different chemical base compounds available,[29] and the redox-flow battery technology, two types emerge as probable for an larger-scale inclusion in the power grid. Lithium batteries have an energy density of $w = 110..190\,\mathrm{Wh/kg}$ and a price range of $170\,\text{€/kWh}$ to $600\,\text{€/kWh}$, but need to be replaced after 400 to 1900 loading cycles. Natrium batteries are high-temperature batteries, featuring $w = 100..165\,\mathrm{Wh/kg}$ at $265\,\text{€/kWh}$ to $645\,\text{€/kWh}$, and have a longer life span of 2500 to 8250 cycles. The efficiency of natrium batteries ($\eta = 72..81\,\%$) is, however, lower than that of their lithium counterparts ($\eta = 90..97\,\%$). Batteries are well suited to provide controlling power, but not to level renewable energies such as wind power. For example, a battery park in Germany containing 25 000 lithium-ion batteries can supply 5 MW of power (Deutsche Presseagentur, 2014).

The aforementioned storage technologies must be considered 'short-term,' except for the hydroelectric power plant with pump-filled reservoir. Technologies that can provide greater amounts of power storage are categorized with the term *power to gas*. Electric power is used to generate and store a form of gas

---

[28]Based on $C = \varepsilon \frac{A}{d}$

[29]Lead-acid, nickel, lithium, natrium

that in turn can used to drive a power plant. Useful gases are methane and hydrogen. Both can be stored in caverns or, in the case of methane, also in pore storage with energy densities of 14 300 Wh/kg, 34 000 Wh/kg and 13 450 Wh/kg. Efficiency, however, does not depend on the storage technology in use alone. In order to provide a full electricity–gas–electricity circle, a transformer, an electrolyzer, a means to methanate, a compressor, a type of storage, and finally, a kind of discharger must be used. Each one of these elements has its own efficiency. Thus, using hydrogen yields an efficiency of $\eta = 34..51\%$ and methane of $\eta = 30..38\%$.[30] How much power a power-to-gas facility uses when storing depends on the electrolyzer; when feeding, the values of a gas turbine power plant apply.

Power Grid Infrastructure

The most obvious sign of a power grid are its cables, which are typically conducted as overland power lines. Their form is chosen to accommodate the voltages of the lines they carry, which, in turn, hew to the distance the cable spans.[31] Several nominal voltage values exist; Table 2.1 lists those most often encountered. The highest voltage, which operation is technically and economically feasible, is 1500 kV (Flosdorff and Hilgarth, 2005).

The grid thus has been organized as a hierarchy: the *transmission grid* carries the highest voltages, called *extra-high voltage*, and is used to transmit power over large distances. Most traditional power plants feed into the grid at high voltages since centralized power generation means that the generated power needs to be transmitted over larger distances in order to reach consumers.

The transmission grid feeds into the *distribution network* through transformers. At lower voltages that are now called *high voltages*, this part of the grid distributes power to large customers as well as to towns or parts of a city. NB. that wind farms and photovoltaic power plants feed into the distribution network; their conceptionally lower rated power output makes a connection to the transmission network more sensible.

Medium voltages are often encountered within cities or used to connect industrial customers. These voltages still range from 3 kV to 30 kV; only the so-called *low voltage* reaches private consumers—known as *secondary customers*—with 230 V to 400 V.

---

[30]Cf. Sterner and Stadler (2014, Tab. 8.30).

[31]$P_{diss} = I^2 R = (\frac{P}{U\cos\phi})^2$ designates the dissipated power, which decreases by voltage squared.

The resistances encountered in the power grid's cabling depend not only on the length of the cable itself, but also on the material it is made of and the configuration in which it is deployed, along with the temperatures encountered in the specific configuration. Reinforced conductors also experience eddy currents, and thus, a theoretical approach is often eschewed in favor of empirical data (Heuck et al., 2010, p. 233). Reference works typically offer tables containing the usual values for different cables and landlines in the usual configurations.[32]

No part of the power grid infrastructure is neutral with regards to the power transmitted, but dissipates power. This loss is composed of three parts:

1. Voltage-dependent loss

2. Current-dependent loss

3. Compensation loss.

Voltage-dependent loss occurs as soon as the line carries a voltage. It is caused by the fact that no insulation is, in reality, ideal. This loss is expressed by:

$$P_{VU} = n \cdot G' \cdot l \cdot U^2 \, , \tag{2.1}$$

where n is the number of parallel cable systems, G' is the conductance per unit length in Siemens per meter, l is the length of the cable, and U is the voltage.

While the voltage-dependent loss changes only with the voltage, the current-dependent loss changes with the load of the cable:

$$P_{VI} = \frac{1}{n} R' \cdot l \left(\frac{S}{U} \right)^2 \, . \tag{2.2}$$

Here, R' expresses the resistivity in Ohm per meter, and S denotes the *apparent power* transmitted.

Compensation losses occur in cables with reactive power compensation, which is necessary on 380 kV land lines starting from 20 km. Compensation loss is computed to:

$$P_{VK} = n \cdot (1 - g) \cdot k \cdot Q'_C \cdot l \, , \tag{2.3}$$

with Q'_C being the capacity of the cable, g the Q factor of the compensation inductor, and k being the inductor's compensation level.

---

[32]Cf. Heuck et al. (2010, pp. 742) and Oeding and Oswald (2011, A.11 to A.14).

The combination of the three terms forms the total power dissipated, which every distribution system must try to minimize. This can be achieved either by design—for example, landlines have lower resistance values than earth cables—, or by simply reducing the distance power has to travel. E.g., a typical 380 kV land line with bundle conductor,[33] designed for 1.1 GW per three-phase system, dissipates about 1 % of power transmitted per 100 km at maximum load (Oswald, 2007).

Each part of the power grid has a maximum capacity in terms of power it can be loaded with. The specific resistance of the material, varies not only with the material itself, but also with its temperature. The more power is transmitted via a cable, the more it heats up, finally reaching its limit. The same effect defines the capacity of all other parts of the power grid, espcially transformers. An accepted rule of thumb is that life expectancy for insulation in transformers[34] is halved for every 7 °C to 10 °C increase in operating temperature (Walling and Shattuck, 2007). They require cooling,[35] which defines their load limit.

Highest power flows occur in short-circuit situations, which can happen for numerous reasons: the insulation can be worn down due to constant overload,[36] because of cables coming into contact with each other, caused by, e.g., wind, or due to cables coming into contact with the ground, because of a tree toppling over, or a cable that sags too much because of its load.[37] The short-circuit power flowing then excessively exceeds the normal power, causing more heating over a short span of time and therefore imminent damage.

Short-circuit capacity is therefore an important rated value for parts of the power grid. It quantifies the maximum stress of an electrical unit and the breaking capacity of a circuit breaker.[38] The required short-circuit capacity rises with development of the power grid.

One might expect that measurements of the power grid's state are readily available since they form the variant, time-dependent counterpart to the infrastructure's static parameters. However, this is not the case: often, parts of the power grid are not equipped with sensors, because all relevant pieces

---

[33] 4×564/72 Al/St

[34] More generally, all electric machines

[35] Larger transformers are cooled with dielectric, i.e., non-conducting liquids, often transformer oil.

[36] This can, in a broader perspective, be a sign of an overstrained power grid (Berg and Fritze, 2011) and leads to burning transformers (Fleischhauer and Nelles, 2007; Seller and Röderer, 2015).

[37] The cable heats, as previously described; the heating causes the material to expand and thus the cable to grow longer, which leads to it sagging.

[38] The interested reader is referred to Flosdorff and Hilgarth (2005, pp. 142) for details.

of infrastructure are generously dimensioned for not only for peak load, but even for an outage of another part. Then, the remaining ones would have to shoulder the load of two. This design principle is called the *n-1 criterion*: if, for the operation of a system, n objects are available, and the loss of one object does not impact the system's operation as a whole, the n-1 criterion is complied with. The German power grid is designed according to the n-1 criterion (Berndt et al., 2007).

However, considering the smart grid, in which active and reactive power will be traded on a short-term basis due to volatile power generation, it is important to estimate the impact of this variable load-behavior on the grid. Power flow analysis exists as a mathematic approach to this problem. To estimate the power grid's future state given the initial state and known changes on load and generator buses. It is a non-linear problem that can be solved computationally; details can be found in Powell (2005) and Momoh (2012, Chapters 3 and 4).

Recalling Fig. 2.1, one can easily see that the classic power grid implements a hierarchy: power is generated in a centralized manner with few large power plants feeding 500 MW to 1500 MW[39] into the power grid. From the transmission network that is meant to cover large distances, power is stepped down into the distribution network, whose purpose is described by its telling name. The customer is on the receiving end; thus, the traditional power grid resembled a hierarchical tree structure that is ever more finely ramified. Table 2.1 strengthens this point (Flosdorff and Hilgarth, 2005).

However, PV installations on roofs naturally feed into the low voltage power grid, whilst larger photovoltaic power plants feed into the distribution network, where also wind farms are connected. A transformer is a two-way device; on exceptionally windy or sunny days, power actually flows from the distribution network back into the transmission network. This had originally never been considered since power always flows from generator to consumer and power generation in lower voltage grids that could not only satisfy but exceed the demand of an area fed by a distribution network was practically unthinkable. Renewable energy sources thus constitute a paradigm shift in generation, as well as in distribution towards local, decentralized power generation.

Bush (2014) argues that the more the power grid becomes a true mesh instead of a (cyclic)[40] tree structure, the more it can be seen analogous to computer networks and compares terms (Bush, 2014, Tab. 1.1). This originates

---

[39]Cf. Table 3.1.

[40]Meshed power grid topologies have long since existed since even a circle structure provides redundancy to allow power transmission even if a line becomes faulty.

Table 2.1: Nominal voltages in the power grid and their application

Nominal Voltage [kV]	Voltage Range	Application
0.23, 0.40	Low Voltage	Small Consumer
3, 6, 10, 15, 20, 30	Medium Voltage	Industrial Consumer, City Feed-in
60, 110	High Voltage	City/Overland Feed-in
220, 380, 500, 700, 735	Extra-High Voltage	Metropolitan Area Feed-in, Collective Economy

from the fact that sensors providing detailed real-time data of the power grid's state are mostly not present because they were not needed in the classical grid (Calpe, 2015). But Bush develops this thought further: He goes so far as to state the emergence of a "power system information theory" that "explicitly combines power systems and information theory," noting that "while a unification of Maxwell's equations and Shannon information theory is suggested for a power system information theory, the unification could perhaps more easily take place at a simple level, such as Kirchhoff's laws [...]" (Bush, 2014, pp. 186, 189).

From the thought experiment of Maxwell's Daemon (Knott, 1911; Maxwell, 2011) and Landauer's Principle (Landauer, 1961; Bennett, 2003), we know that the energy released by an irreversible logic operation—i.e., a loss of information entropy—is

$$E = \mathrm{k}T \ln 2 \ , \tag{2.4}$$

where k is the Boltzmann Constant and T the absolute temperature of the system. Bush further explains the correlation between energy and a bit transmitted. The energy required to transmit one bit given an infinite amount of time is shown by:

$$E_b^{min} = N_0 \lim_{C \to 0} \frac{2^C - 1}{C} = N_0 \ln 2 \ , \tag{2.5}$$

where N_0 is the noise variance $N_0 = E[nn*]$ of a complex Gaussian distribution and C is the information channel's capacity.[41] Bush then states that, given the

---

[41] A communication channel's capacity can be expressed as $C = B \log_2(1 + \frac{S}{N})$, where B denotes the bandwidth, S the signal, and N the noise of the channel.

difference in a power grid between power generated and power received, ΔE, it will take an amount of bits equal to

$$D = \frac{\Delta E}{2\mathrm{k}T \ln 2} \qquad (2.6)$$

in order to compensate for this power loss by a smart operation of the power grid (Bush, 2014, Chapters 6.2 and 6.3).

We can therefore conclude that the power grid as infrastructure, although seemingly defined a set of parameters, forms in its mapping to a communication network the vital backbone of the development of a new organization of power generation and power consumption. Thus, "bits (of communication) per kilowatt (of active power delivered)," and "maximum transmission efficiency (power delivered/generated) per bit (of communication)" (Bush, 2014, p. 201) form two important metrics that can be used to measure a system that uses digital information distribution and processing to increase the efficiency of the power grid.

This thesis has hitherto used the term *smart grid* in a manner that suggested that its meaning is immediately obvious to the reader. Bush has tried to follow the historic trail in order to find the first usage of the term (Bush, 2014, Section 1.5), tracing it to a time after the Northeast Blackout of 2003. As for a definition, this work will use a part of the US code from 2007 that stipulates "[the smart grid makes] [i]ncreased use of digital information and controls technology to improve reliability, security, and efficiency of the electric grid" (Bush, 2014, ibid.).

Power Consumption

In the power grid, due to its alternating current nature, two types of power exist. The apparent power, expressed through $S = I_{rms}V_{rms}$, consists of *active power* and *reactive power*. Active and reactive power can be expressed as vectors; thus, the Pythagorean Triangle equation is applicable and yields:

$$S^2 = P^2 + Q^2 \; . \qquad (2.7)$$

All three are products of voltage and current, which is expressed as watts; in order to distinguish between them, only active power is expressed in terms of watts, while apparent power has the unit *VA* (read: volt-amperes) and reactive power is expressed as *VAr*, volt-amperes reactive. Reactive power is power that is not used for real work, but exists when current and voltage are not in perfect alignment. It is power that oscillates between electric and magnetic fields in the power system:

$$Q = I_{rms} V_{rms} \sin \Phi \; . \tag{2.8}$$

Reactive power is not desired, but is introduced as a natural consequence of inductive loads that shift the phase angle $\Phi = \angle IV$. Compensating reactive power is an ancillary service and necessary as most non-resistive loads in the power grid are inductive; large factories are typically billed not only for active power like private customers are, but also for the reactive power they introduce in the system. Condenser batteries are used to 'counter' reactive power.[42] A power plant can supply the ancillary service by exciting of their generators.

The traditional grid already relied heavily on planing; power consumption is estimated using *Standard Load Profiles* (SLP). For Germany, they are computed by the *Bundesverband der Energie- und Wasserwirtschaft e. V.* (BDEW).[43] Fig. 2.6 shows a historic load profile from New England in 1919 (National Museum of American History, 1919). In the modern grid, these load profiles are either a mapping of the corresponding consumer—e.g., household, bakery, etc.—, or have huge deviations due to photovoltaic installations on rooftops that change the overall consumption/production values and make the consumer a *prosumer*, i.e., a producer and consumer in a single entity. Forecasting algorithms need to accommodate these situations.

The SLP assume a consumption behavior that is insensitive to the grid's state. In the future, a consumer might be incentivized to reacting to signals coming from the power grid. Especially industry consumers can help shifting loads: They can consume surplus power when available and throttle consumption at a later point when power is in shorter supply. Cold storage houses, for example, can act as thermal energy storage in using surplus power for cooling well beyond the normally required temperature and switching off the chiller during a peak load situation.

2.2   Simulation and Modeling

Simulation is "the process of designing a model of a real system and conducting experiments with this model for the purpose of understanding the behavior of the system and/or evaluating various strategies for the operation of the

---

[42]A capacitor shifts the phase angle $\Phi = \angle IV$ by 90°, i.e., the AC voltage lags the current: $I_C = I_0 \cos(\omega t + 90°)$. In contrast, inductive loads shift the phase angle by a negative value.

[43]En. *German Association of Energy and Water Industries*

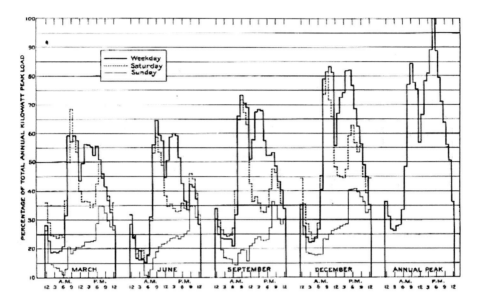

Figure 2.6: Loads of utilities in the Eastern New England Division in 1919

system" (Shannon, 1998). According to Banks et al. (2013), a simulation is appropriate when the given models and their interactions need to be observed over time in order to arrive at a result, while the complexity of the overall system is too high to allow a single, formulated calculation of it. Computer networks especially lend themselves to simulation, for obvious reasons: the nodes exchange messages that influence their behavior greatly while the state of all nodes in the system must be maintained over time; observing this message exchange and the exhibited behavior in different situations, along with the nodes' internal state, is often the main interest of a network researcher.

Simulations of computer networks can be conducted as *discrete-event simulations*. Here, the simulation is driven by discrete events that indicate a change in the simulated environment, such as the transmission or reception of a message. Timekeeping is done in an abstract manner based on *ticks*. A tick[44] marks the occurrence of one or more events, but is not related to a particular date or time. If an event is associated with a real-world time, the timespan between two ticks can be anything from the fraction of a second to many days, months, or years.

---

[44]Inspired by the tick-sound of a clock

Discrete-event simulations are not only useful for the simulation of computer networks, but anything that can be modeled through discrete events. They do not require specialized simulation software and simple queueing simulations can even be conducted with a piece of spreadsheet software. For an introduction to the basics and implications of this type of simulation, along with the relevant statistics knowledge, the reader is referred to Kelton and Law (2000); Banks et al. (2013); Fishman (2013).

Software or frameworks for discrete-event simulation exist in plenty. The network simulators NS-2 (Bajaj et al., 1999), NS-3 (The NS-3 Project, 2015), J-Sim (Sobeih et al., 2006), SSFNet (Cowie et al., 1999), JiST/SWANS (Barr et al., 2004), OPNET Modeler (OPNET Technologies, Inc., 2015), and Qualnet (Scalable Network Technologies, 2016) have inspired the design of the OMNeT++ simulation environment (Varga, 2001; Varga and Hornig, 2008).

OMNeT++ defines units in the form of modules in order to structure the model. Modules have gates that send and receive messages; OMNeT++ as a whole relies on message passing to exchange information between modules. These modules can be grouped together to form compound modules, creating hierarchies in the model.

OMNeT++ is written in C++, which above all permits the use of existing libraries written in C/C++ to extend the simulator. This way, real-world implementations of protocols such as TCP can be used directly. This extensibility also makes OMNeT++ modular; the simulation core can be embedded into other applications. In addition to that, it features an *Integrated Development Environment* (IDE), wherein the user can set up the simulation, write C++ code in order to define the behavior of its own modules, and view results recorded in previous simulation runs.

The definition of the simulation environment's layout as well as the simulation parameters is done in a description language specific to OMNeT++, called *Network Description* (NED).

While the software framework delivers the engine of the simulation, the most important part of any simulation are its models. Any modeller needs to create a *conceptual model* of what he or she is about to simulate before everything else. A conceptual model, according to Robinson (2004), is the process in which "[t]he modeller, along with the clients, must determine the appropriate scope and level of detail of model [...]" and determines its effectiveness (Law, 1991). Authors have identified several qualities of an effective model (Willemain, 1994; Brooks and Tobias, 1996), from which Robinson (2004, Chapter 5.4.1) syndicates the "four main requirements of a conceptional model: validity, credibility, utility, and feasibility."

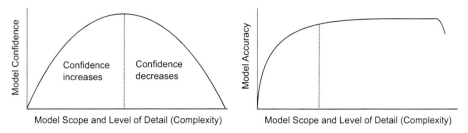

Figure 2.7: Model complexity versus model confidence and model accuracy

Validity The conceptual model is, according to the modeller, sufficiently accurate for the purpose at hand.

Credibility The conceptual model is, according to the client, sufficiently accurate for the purpose at hand.

Utility The conceptual model is, according to both, the modeller and the client, useful in order to aid decision-making within the specified context.

Feasibility The conceptual model can according to both the modeller and the client, developed into a computer model.

The general goal is thus to keep the model as simple as possible while still meeting the objectives of the simulation study. Simple models can be developed faster, are more flexible and typically require less data while still being accurate enough. Most importantly, their structure is better understood (Innis and Rexstad, 1983; Ward, 1989; Salt, 1993; Chwif et al., 2000). Fig. 2.7 shows that not only will the confidence in the model decrease with increasing complexity, but also there will be little gain when increasing the model's complexity past a certain point; it might even become less accurate.[45]

Complex models can be simplified in two ways: either by removing components that have little to no impact in the model's accuracy, or by replacing parts of the model with simpler ones while still maintaining a satisfactory level of accuracy. Innis and Rexstad (1983); Courtois (1985) discuss methods of model simplification; Pidd (1999) argues the reverse: that one should start with the simplest model possible and add to its level of detail step by step until the desired accuracy is reached.

---

[45]Based on Lobao and Porto (1997); Robinson (1994, 2004)

Methods of model simplification include grouping of entities, black-box modeling and the replacement of components with random numbers. When entities are grouped, the modeller describes a whole group by an aggregated model instead of modeling individuals one by one. This can be seen analogous to the flyweight pattern in software design (Gamma et al., 1995a). When the designer choses black-box modeling, he does not try to model the intrinsic workings of an entity, but merely tries to represent its responses given a set of inputs: the entity is conceived as a black box whose inner workings are unknown. Instead, it can be represented as a function. Artificial neural networks lend themselves to this kind of modeling regarding more complex components since they can learn to represent any function.[46] Finally, instead of creating a model by describing the entity's behavior, the designer can choose to represent the particular component by random numbers by choosing a random number distribution with parameters matching the entity's observed values. A gentle introduction to the required statistics knowledge, the process of selecting the appropriate random number distribution and testing the fitting, including the well-known χ^2 test, can be found in Robinson (2004, Chapter 7).

A conceptual model can be represented in different formats. A *component list* is a textual representation in the form of a table giving the components and noting their respective complexity. In terms of graphical formats, the *Unified Modelling Language* (UML) can be used to represent a conceptual model (Richter and Marz, 2000; Knaak and Page, 2006).

The most important property of a conceptual model is its objective: "A model has little intrinsic value unless it is used to aid decision-making [...]" (Robinson, 2004, p. 80). It is therefore necessary to define the metrics with which its success can be measured.[47] In order to indicate success or failure of a simulation run, inputs and outputs need to be defined. In almost all situations, this means numerical and graphical reports. Ehrenberg (1999) discusses the merits of graphical and numerical reports.

It is strongly advisable to decouple the definition of input and output values and sources, and, if possible, their analysis from the simulation software itself. I.e., the *separation of concerns* (Dijkstra, 1982; Reade, 1989) should be followed. This allows the modeller to reason about simulation parameters and the response of the system without having to write a program; he can focus on the task at hand, potentially with the customer, who, in most cases, is not apt to programming. OMNeT++ uses the NED (Varga and Hornig, 2008) for that

---

[46]Cf. Section 2.4.

[47]The metrics defined in Bush (2014) lend themselves exactly to that purpose.

purpose; a more generalized approach is *Simulation Experiment Specification via a Scala Layer* (SESSL) (Ewald and Uhrmacher, 2014, 2012).

2.3 Computer Networks

The most important foundation of today's communication networks, especially the Internet, is the *International Standards Organization* (ISO)/*Open Systems Interconnection* (OSI) reference model (Zimmermann, 1980), also colloquially known as the 'ISO/OSI stack.' It consists of seven layers in total, as shown in Fig. 2.8.[48] Each layer in the ISO/OSI stack model has its own purpose, independent from the others; no task is shared between any two layers. At least in theory, this also entails modularity: One can exchange protocols at one layer without having to touch any of the other protocols on the other layers; the operation of each layer in the OSI reference model is completely transparent. When the user browses a web site, the *Hypertext Transfer Protocol* (HTTP) that is used to transfer the site's content does not need any information about the underlying link, whether it is a wireless LAN connection or established using a cable. For this, the OSI reference model uses the concept of encapsulation: to a protocol in a certain layer, everything the upper layers produce is simply payload. Consequently, it has no knowledge of the operational data of the lower levels' protocols.

Although numerous people have argued that, in the face of today's development of the Internet on a large scale, the OSI reference model is obsolete[49] and have therefore offered different approaches (O'Malley and Peterson, 1992; Zitterbart et al., 1993; Handley, 2006; Reuther and Henrici, 2008; Henke et al., 2010), it still forms the centerpiece of modern communication system architectures.

Layers 1 and 2 define access to a physical medium and how communication across a single link is to be conducted. The ubiquitous Ethernet (Institute of Electrical and Electronics Engineers (IEEE), 2012b) and Wi-Fi (Institute of Electrical and Electronics Engineers (IEEE), 2012a) protocols reside in these two layers, along with those used in cellular networks, i.e., *Global System for Mobile Communications* (GSM) (Redl et al., 1998), *Universal Mobile Telecommunications System* (UMTS) (Kreher and Ruedebusch, 2007), *High Speed Downlink Packet Access* (HSDPA) (Holma and Toskala, 2007), and *Long-Term Evolution* (LTE) (Sesia et al., 2009).

---

[48]Based on (Fall and Stevens, 2012, p. 9)

[49]Indeed, layer violations are widely known, such as the combination of link and network layer in the *Dynamic Host Configuration Protocol, version 4* (DHCPv4) (Droms, 1999).

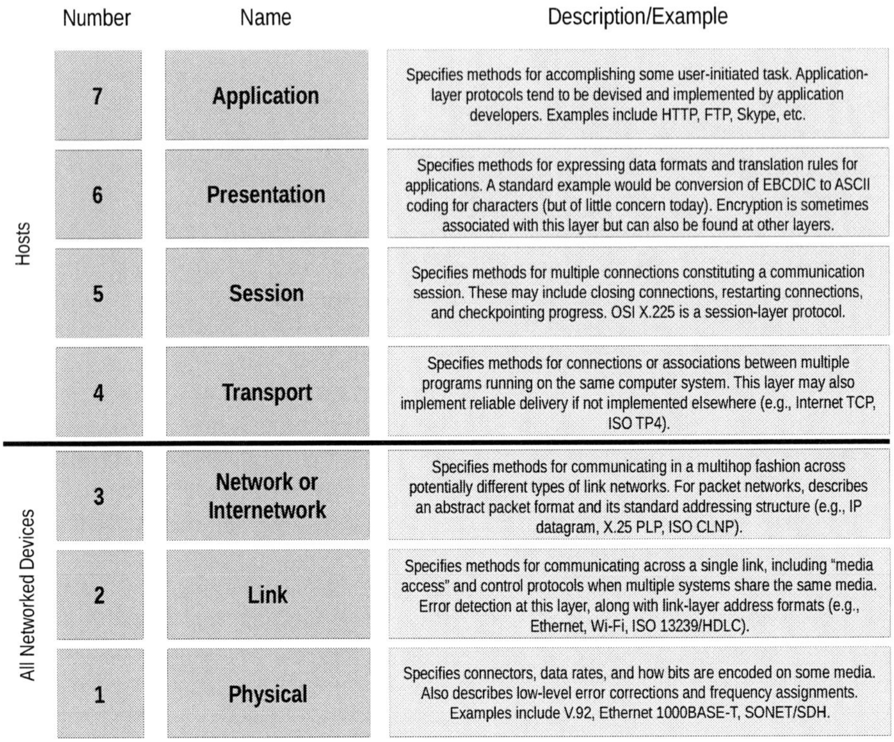

Number	Name	Description/Example
7	**Application**	Specifies methods for accomplishing some user-initiated task. Application-layer protocols tend to be devised and implemented by application developers. Examples include HTTP, FTP, Skype, etc.
6	**Presentation**	Specifies methods for expressing data formats and translation rules for applications. A standard example would be conversion of EBCDIC to ASCII coding for characters (but of little concern today). Encryption is sometimes associated with this layer but can also be found at other layers.
5	**Session**	Specifies methods for multiple connections constituting a communication session. These may include closing connections, restarting connections, and checkpointing progress. OSI X.225 is a session-layer protocol.
4	**Transport**	Specifies methods for connections or associations between multiple programs running on the same computer system. This layer may also implement reliable delivery if not implemented elsewhere (e.g., Internet TCP, ISO TP4).
3	**Network or Internetwork**	Specifies methods for communicating in a multihop fashion across potentially different types of link networks. For packet networks, describes an abstract packet format and its standard addressing structure (e.g., IP datagram, X.25 PLP, ISO CLNP).
2	**Link**	Specifies methods for communicating across a single link, including "media access" and control protocols when multiple systems share the same media. Error detection at this layer, along with link-layer address formats (e.g., Ethernet, Wi-Fi, ISO 13239/HDLC).
1	**Physical**	Specifies connectors, data rates, and how bits are encoded on some media. Also describes low-level error corrections and frequency assignments. Examples include V.92, Ethernet 1000BASE-T, SONET/SDH.

Layers 4–7 are labelled *Hosts*; layers 1–3 are labelled *All Networked Devices*.

Figure 2.8: The ISO/OSI reference model

Layer 3 is concerned with routing: Protocols in this layer span link networks, potentially of different types. They define the address format of *hosts*[50] and perform multiplexing by defining networks as packet-switched. The most commonly known representatives are the older, but seemingly still-prevailing *Internet Protocol, version 4* (IPv4) (Postel, 1981a) and its newer counterpart, the *Internet Protocol, version 6* (IPv6) (Deering and Hinden, 1998). The latter one can be deemed of high importance for the smart grid: The number of nodes participating in a grid-wide communication network will be significant and the IPv4 address space of 2^{32} addresses is already exhausted (Cannon, 2010; Lee et al., 2011). IPv6, in contrast, offers 2^{128} addresses.

Since computer networks can be treated uniformly, i.e., independent from

---

[50]NB. that a host is a logical concept, and can be made of several physical interfaces.

the underlying transmission technology, as a graph theory problem, principles
from that domain form the rationale underlying today's routing algorithms. The
Internet is subdivided into *Autonomous Systems* (AS). The ASs communicate
their routing information using an *Exterior Gateway Protocol* (EGP); the most
commonly used EGP is the *Border Gateway Protocol* (BGP) (Rekhter and Li,
1995). It is a path vector protocol: For each node, the router stores a vector
containing a concatenation of all edges, i.e., paths a packet needs to travel in
order to reach the designated node. Routers transmit their vector database to
their neighbors in order to allow them to build their own. BGP, because of its
nature as a path vector protocol, has a fast convergence.

Within an AS, the network administrator is free to choose whatever routing
protocol he likes. Often, *Open Shortest Path First* (OSPF) (Coltun et al., 2008)
is chosen for the task. It is a link-state protocol that builds a sink-tree for each
router, based on *Dijkstra's Algorithm* (Dijkstra, 1959).

Routing algorithms are important to understand how a packet or datagram
finds its way from sender to receiver, the basis for any point-to-point communication in a computer network. They are also applicable to the power grid:
Chapter 6 shows how. The details of the actual protocols, be it BGP, OSPF, or
other, are discussed at length in various literature, e.g., by Tanenbaum (2003).

Most packets in IP networks travel point to point; they are sent in a unicast
fashion from one sender to one receiver. Multicast, i.e., a packet sent by one
transmitter to several receivers, needs to be implemented in an efficient fashion—
a sender does not simply transmit n packets to n receivers. Instead, the packet's
destination address is a multicast address, i.e., a special address, that belongs to
a particular multicast group. Groups can be formed on different bases; compare
Haberman (2002); Savola (2011) for details.

Layer 4 of the OSI reference model harbors all transport protocols. Their
responsibility is to transfer between programs.[51] The most commonly known
protocols here are the *User Datagram Protocol* (UDP) (Postel, 1980) and the
Transmission Control Protocol (TCP) (Postel, 1981b).

The two differ in their intended goals and thus mode of operation: The
TCP provides the concept of a connection between two programs within which
a stream of data exists. TCP preserves the order of data transmitted and
minimizes[52] loss of data through explicit segment acknowledgement and re-

---

[51] Notice the hierarchy: Layer 2 is responsible for transferring data between adapters on a
single link, Layer 3 transfers packets between hosts in networks, and Layer 4 finally reaches
applications on hosts.

[52] No algorithm can guarantee complete protection against loss of data: A link may become
offline, e.g., because of a hardware failure.

transmission of all data lost. En lieu of this, the TCP also has mechanisms for congestion control, i.e., for throttling the rate of transmission in order to neither overload the receiver nor the network.

This comes at a cost: While the TCP can reliably saturate a link with a stream, retransmission introduces latency whenever segments are lost. This is the *raison d'être* for the UDP, which cares neither for the order of datagrams or for loss minimization or congestion control. Through this 'bare-bones' design the application programmer needs to concern herself only with the network's latency; she exerts direct control over the transmission and reception behavior of her program. This is especially important for low-latency applications like *Voice over IP* (VoIP) or video streaming. If mechanisms such as rate throttling or retransmission are desired, the application programmer needs to implement those on top of the UDP.

Higher levels define application protocols, incorporate encryption and authentication mechanisms, and also concern themselves with the encoding of user data.

If a person wants to avoid binary encoding of data that would be the most efficient way of representing data in terms of encoding/decoding speed and space consumption in order to achieve human-readability, two textual formats are well established: The *Extensible Markup Language* (XML) (Bray et al., 1998) and the *JavaScript Object Notation* (JSON) (Bray, 2014). The JSON can be chosen over the XML for the faster performance of parsers and generators reading respectively writing the format, fewer space requirements and arguably better readability over the XML, even though it offers a reduced feature set compared to its counterpart (Nurseitov et al., 2009).

The emerging *Common Information Model* (CIM) (International Standards Organization (ISO), 2005) models objects of transmission, distribution, and generation of electric power and aims at providing a common data exchange format for every aspect in the power grid, including trade. It uses the XML due to its complexity to benefit from the namespacing XML offers. However, the smart grid knows many participants with difficult connectivity such as remote wind farms or smart meters in cities, where a robust technology is desirable. Long wave-based protocols such as LoRa (LoRa® Alliance, 2016) can fulfill these requirements, but offer low data rates, e.g., 0.3 kbit/s to 50 kbit/s due to the underlying physics, which make general deployment of the rather heavyweight CIM a point of discussion.

A computer network, especially considering the Internet, has many points at which an attacker could intercept datagrams to learn secrets, or worse, modify the data transmitted; security is thus paramount. The abstract term 'security' consists of two distinct tasks: The most obvious task is to encrypt data so that no entity can 'eavesdrop' on a communication between two parties; also, any corruption of the transmission needs to be detected. Additionally, the parties require a means to identify themselves to each other: Both end points need to ascertain that their counterpart is actually the desired communication partner and not a man in the middle.

A user typically knows the underlying protocols indirectly from secured web connections; the basis for this is the *Transport Layer Security* (TLS) suite (Dierks and Rescorla, 2008) that assembles encryption and authentication. The latter is achieved using certificates. Each party that wants to authenticate itself presents a certificate, which not only carries that party's common name, but also a validity date and a signature. This signature is affixed by the certificate's issuer that acts as a *trusted third party*: Both communicating parties trust the issuer and can thus, if the issuer authenticates the respective certificates, trust each other. TLS itself is used to provide security to many protocols; the underlying mechanisms are also used in another protocol that provides security in IP networks: *Internet Protocol Security* (IPsec) (Kent and Seo, 2005).

Providing a detailed introduction of security measures available for computer networks and possible vectors of attack is beyond the scope of this work; the interested reader is referred to Tanenbaum (2003, Chapter 8) for a gentle introduction and Stallings (2013) for a broader lecture regarding cryptography and its application in computer networks.

A full protocol stack as hitherto described can prove to be resource intensive, especially considering embedded devices for sensor hardware, even with stripped-down network stack implementations (Sehgal et al., 2012). Due to the OSI reference model's modularity, the protocols mentioned in the previous paragraphs are not the only ones that provide certain features such as transport reliability and security. For exactly that reason Kim et al. (2011) propose the *Scalable and Secure Transport Protocol* (SSTP) for smart grid devices. It resides on Layer 4 of the OSI reference model, and provides reliable delivery through and acknowledgement/retransmission mechanism similar to the TCP. In contrast to the latter, however, it is not meant for high-volume data transmissions, and therefore features no congestion control. Instead, a number of low-volume messages are exchanged, which also explains the absence of an order-preservation mechanism. It pays attention to scalability and strives to reduce latency. The SSTP also includes security features instead of relying on TLS. While this approach saves resources, it can also quickly form a possible vector of attack as the *Open Smart Grid Protocol* (OSGP) has shown, as discussed below.

It should be noted that in their paper, Kim et al. (2011) perform a simulation to compare the SSTP with the TCP, using the TCP Reno congestion control mechanism. Although published in 2011, newer methods of TCP congestion control, such as CUBIC (Ha et al., 2008), have already become available. This, however, does not invalidate their motivation and approach, and makes the SSTP a worthwhile consideration.

The hitherto complete description of the OSI reference model allows us now to understand the concept of an *overlay network*: An overlay network constitutes a logic network on existing infrastructure; it uses an already existing network architecture and topology to form its own. Overlay networks often implement their own addressing and routing schemes independent from the underlying communication architecture, thus forming their own topology. The overlay network concept is used as a basis for the protocol presented in Chapter 6. Other existing overlay networks are *peer-to-peer networks* such as Chord (Stoica et al., 2003).

While the OSI reference model and its most commonly known implementation, the Internet, is ubiquitous, it has not ousted other approaches to networking or their perceived necessity. In fact, a widespread protocol dealing with smart grid devices, the OSGP (ETSI, 2012), is not based on the OSI reference model and the protocols adhering to it, but on the *ISO/International Electrotechnical Commission (IEC) 14908 Control Networking Standard For Smart Grid Applications* (International Standards Organization (ISO), 2012).

The OSGP consists of a number of protocols; the ISO/IEC 14908 communication standard also defines a protocol stack model. The OSGP uses its own addressing scheme and routing protocol. It is optimized for the interaction with smart grid devices, such as smart meters, and aims at providing an efficient mechanism for data querying and control commands. To that end, the OSGP also defines its own authentication, authorization, and encryption layer.

Specifying technologies similar to those already in use, especially in terms of security, opens the protocol up to attack similar to those seen on the Internet, especially regarding weak cryptographic cyphers, as Jovanovic and Neves (2015) have shown.

2.4   Artificial Intelligence

Agent Concept

Russel and Norvig (2010) describe agents as "anything that can be viewed as perceiving its environment through sensors and acting upon that environment through actuators" (Russel and Norvig, 2010, p. 10). The term *agent* originates from Turing's (1950) famous paper, "Computing Machinery and Intelligence."

The agent's behavior is described by the agent function that selects an action based on the agent's internal state, its knowledge, and the input data gathered by its sensors. A program implements this agent function. The terms *sensor* and *actuator* describe any device, hard- and software alike, that is—in case of the sensor—able to provide the agent function and thus the implementing software with information about its environment in order to update the agent's internal state, and to—in case of the actuator—influence its environment through actions chosen based on the agent's sensory input, internal state, and agent function. The agent function chooses an action from the pool of actions that are available to the agent in order to maximize the agent's performance in regards to its defined goal.

The simplest form of an agent is purely reflex-based: It exhibits an action based on the environment it perceives at that particular moment; it only reacts. Such a design is called a *simple reflex-based agent*. This simple behavior model has strong conceptional ties with psychological behaviorism (Skinner, 1953). If the world it cannot currently perceive is also part of the action-selecting process, then it must maintain a model of the world, and thus becomes a *model-based reflex agent*.

However, in order to exhibit stringent behavior in the long term, it also needs a definition of its global goal, making it a *goal-based agent*. Furthermore, an agent can possess a function that judges different ways to reach the global goal, a utility function that extends it to a *utility-based agent*. If the agent is able to change its behavior based on previous sequences of inputs, actions, and results, it is able to learn, i.e., it becomes a *learning agent*. Russel and Norvig's (2010) standard monograph covering artificial intelligence describes the design and development of agents based on this typology in further detail. A different typology is offered by Nwana (1996).

Wooldridge and Ciancarini (2001) summarize an agent's four most important properties: Autonomy, i.e., the agent's ability to encapsulate a state (or state estimation) and act based on that; reactivity, meaning the agent's perception of its environment, allowing the agent to react to changes in it promptly; proactiveness, since the agent is able to show initiative; and the agent's social ability, as it communicates with other agents, and possibly humans. Strictly speaking, agents cannot be understood in terms of functional systems as they change their environment during execution.

One of the most notable formal approaches to agents, their communication, and how it can generally be used to solve problems in a distributed way, is the

Contract Net Protocol by Smith (1980). Indeed, one can easily identify the ideas in Chapters 4 and 6 due to the general nature of Smith's approach. On closer inspection, the reader will note the differences due to the task at hand, especially in the agents' communication behavior that is dictated by the protocol presented in this thesis. Other approaches to agent-oriented software engineering, i.e., how to model agents, are presented and summarized by Wooldridge and Ciancarini (2001).

From the example of Smith's paper and the characteristics offered by Wooldridge and Ciancarini it becomes obvious that, when designing agents, one does not only need to concern oneself with the software itself, but also with a communication protocol the agents utilize in order to be able to exhibit social behavior. As we know from Section 2.3, communication protocols encompass not only a means of encoding and representing data, but also a behavior that needs to be followed by all communication partners lest the information interchange fails. There is no 'gold-standard' for agent communication, but standard frameworks exist, such as *Java Agent Development Framework* (JADE) (Bellifemine et al., 2007). One can argue that behavioral rules form the most important portion of any agent communication protocol, not encoding, and thus already existing protocols based on standards should be employed as the basis for agent communication. Section 2.3 has listed some, also in the context of the smart grid; the IEC 61850 and IEC 61499 also lend themselves to infrastructure automation in the smart grid (Vyatkin et al., 2010b).

Agents have been used widely in many fields of application, for example, as agents acting in the economy (Cliff, 1997; Greenwald and Kephart, 1999; Kephart et al., 2000; Kephart, 2002; Padovan et al., 2002; Koritarov, 2004; Vale et al., 2011), or for data mining (Cao et al., 2009). In terms of the smart grid, the agent design suggests itself readily and therefore finds publication in different approaches.

McArthur et al. (2007a,b) outline the *Multi-Agent System* (MAS) approach to power engineering applications in two papers presented by the IEEE Power Engineering Society's MAS Working Group. From the idea of self-healing shipboard systems on board combat ships, where power flow needs to be reconfigured in the face of damages on the system during a battle to keep the ship operational (Butler et al., 1999), *self-healing* in terms of dynamic reconfiguration to mitigate the effects of a fault has been defined as a task for the smart grid that can be achieved using software agents (Nagata and Sasaki, 2002; Davidson et al., 2006; Vyatkin et al., 2010a; Zhabelova and Vyatkin, 2011; Higgins et al., 2011).

But not only recovery, but also optimization of the power grid has been targeted by researchers using the MAS paradigm. In a series of papers, Rogers et al. (2010); Aquino-Lugo and Overbye (2010); Aquino-Lugo et al. (2011) use software agents to optimize power flow, for voltage support. Their approach to manage active and reactive power uses an hierarchical chain-of-command structure.[53] In fact, Pipattanasomporn et al. (2009) explicitly state that "[t]he idea behind any multi-agent system is to break down a complex problem handled by a single entity—a centralized system—into smaller simpler problems handled by several entities—a distributed system."

Thus, many designs of agent systems for the smart grid distinguish between different types of agents based on their task, position in the system, or their overall hierarchical design. This Divide-et-Impera approach certainly follows the guidance of good software design, but also leads to very specific, distinct software packages that are tied to particular hardware types, locations, or perform only a single, distinct task, making multiple agents in one installation necessary. The agent design proposed in Chapter 4 does not distinguish between different hardware types in a hard-coded manner.

Machine Learning through Artificial Neural Networks

Artificial Neural Networks (ANN) try to implement machine learning by creating a model of the brain: the neural network. It contains artificial neurons that mimic the behavior of their biological counterparts: a neuron receives a number of stimuli through connections that are summed up and inhibited by a threshold value. It finally activates the neuron, transforming the input into an output. This creation of a structure that is modeled on the human brain dates back to the work of McCulloch and Pitts (1943).

Neurons are connected to each other: one neuron's output forms a part of another neuron's input. These connections are weighted, i.e., values that are 'transferred' between neurons are modified by the connection's weight. Training an ANN describes the task of modifying the neuron's connections' weights such that the network's output matches the desired output given the respective input presented to the ANN.[54] See Fig. 2.9 for a schematic view of a neuron.

A simple form of an ANN is the *perceptron*, invented by Rosenblatt (1957). A perceptron contains neurons that receive input data to the neural network

---

[53]More specifically, the authors organize their agents based on the Incident Command System (U.S. Department of Transportation, Federal Highway Administration, Office of Operations, 2013).

[54]Training also incorporates the modification of a neuron's activation threshold. In order to simplify this, this threshold is externalized in form of a bias neuron.

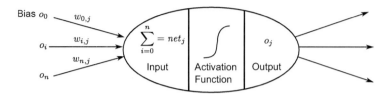

Figure 2.9: Schema of an artificial neuron

and output neurons from which the result of the network's calculation is read. Neurons are grouped into layers according to their position in the network, thus, a perceptron contains at least an input layer and an output layer. Hidden layers are also usual: they are located between the input and the output layer and not visible from the outside, hence the name.

A perceptron is a directed, acyclic structure. Input is fed to the network's input layer neurons, their output travels along connections to neurons in the hidden layers and finally reaches the neurons in the output layer. Perceptrons are thus simple, feed-forward networks. Despite their seemingly simple structure, ANNs can represent any function: A single layer is enough to represent any continuous function, whereas a network with two hidden layers can represent any mathematical function (Cybenko, 1988, 1989). A perceptron that can be trained to represent the XOR function is depicted in Fig. 2.10a.

The works by Cybenko show how important the activation function is and that choosing a sigmoid function is often the most beneficial. Jordan (1995) explicitly discusses the role of the sigmoid function, i.e.,

$$\mathrm{sig}(x) = \frac{1}{1 + e^{-x}} \; . \tag{2.9}$$

The power of an ANN lies in its ability to derive meaningful results from unknown inputs that follow a pattern it has previously been trained to recognize. We can see from the results published by Cybenko that the network's size and layout directly influences its 'memory,' i.e., its ability to learn a pattern. More complex patterns require more complex networks. However, simply increasing the number of neurons and layers is no solution, even if one dismisses memory and computation time considerations: If a network contains too many connections, it will suffer from *overfitting* and loses its ability to generalize, i.e., to derive meaningful values from unknown, but pattern-matching, input values. However, a method called *Optimal Brain Damage* for removing superfluous connections exists (Le Cun et al., 1990; Sietsma and Dow, 1988).

While the perceptron is able to detect static patterns and infer related ones, it has no concept of time. Users often try to work around this shortcoming by creating a vector of items of a time series $x_t, x_{t-1}, \ldots, x_{t-n}$ that is fed as a whole to the perceptron's input layer. While this 'flattens' a time series into a static pattern, it still does not enable the perceptron to learn about a timely series; it merely creates a new pattern.

Researchers have therefore created neural network structures that contain a 'memory.' They introduced an additional layer, called the *context layer*, whose neurons' inputs are the outputs of other neurons in a different layer. The neurons in the context layer save the input they receive and feed it to their associated neuron on their next activation. Thus, each neuron in the context layer feeds the result of the $t - 1$-th run on the t-th activation, acting as a memory. These types of networks are called *Recurrent Neural Networks* (RNN). The most notable works that have introduced this ANN architecture are the papers by Jordan (1986); Elman (1990). The difference between the two approaches is the connection of the context layer: In Elman networks, the context layer is fed by the hidden layer; Jordan networks feed it from the output layer. Fig. 2.10b shows a small Elman RNN.

Hochreiter and Schmidhuber (1997) have extended the RNN's memory through their *Long Short-Term Memory* (LSTM) design. Today, RNNs such as Elman's and Jordan's are called *Simple RNNs*.

In order to train ANNs in a supervised manner, a training set containing a number of samples with known results is presented to the yet untrained network. The training's obvious goal is to reduce the error of the network's output compared with the desired—and known—result of the training sample towards 0. Again, one can borrow inspiration for an algorithm from neuroscience, and follow *Hebb's postulate*:

> "Let us assume that the persistence or repetition of a reverberatory activity (or 'trace') tends to induce lasting cellular changes that add to its stability. [...] When an axon of cell A is near enough to excite a cell B and repeatedly or persistently takes part in firing it, some growth process or metabolic change takes place in one or both cells such that A's efficiency, as one of the cells firing B, is increased." (Hebb, 2012)

The application of Hebb's postulate led to a family of algorithms that use the *back-propagation of error*. Here, the error is propagated backwards from the output through the hidden to the input layer. Since its inception, the original

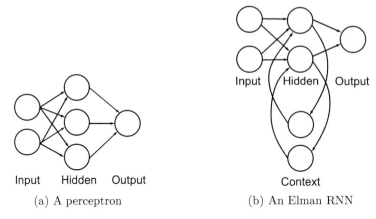

Input   Hidden   Output

(a) A perceptron

Input   Hidden   Output

Context

(b) An Elman RNN

Figure 2.10: Two types of Artificial Neural Networks

algorithm has been modified and revised several times in order to optimize it; the most notable optimization being the Rprop algorithms (Bryson and Ho, 1969; Rumelhart et al., 1986; Riedmiller and Braun, 1992; Riedmiller, 1994a,b; Igel and Hüsken, 2000, 2003; Lalis et al., 2014). RNNs cannot be trained with the original back-propagation algorithm. Instead, those are unfolded over a number of time steps so that they become feed-forward networks. The corresponding algorithm is known as *back-propagation through time* (Mozer, 1989).

In general, finding a good weight configuration for a given neural network is an optimization problem that is NP-complete (Judd, 1990). As a gradient-decent algorithm such as back-propagation of error can get stuck in local minima, other approaches of informed search can be used instead, such as *Simulated Annealing* (SA) (Kirkpatrick et al., 1983), *Particle Swarm Optimization* (PSO) (Kennedy and Eberhart, 1995; Shi and Eberhart, 1998; Clerc, 2012), or evolutionary algorithms (Branke, 1995). Chapter 5 will present a training algorithm in this regard (Ruppert et al., 2014).

Researchers have shown that the number of samples for testing is related to the size of the ANN. Baum and Haussler (1989) state that roughly, given the vector of trainable weights of an ANN, $\boldsymbol{w}$, a number of samples equal to

$$n = |\boldsymbol{w}| \log |\boldsymbol{w}| \qquad (2.10)$$

is necessary; Anthony and Bartlett (2009) offer a more sophisticated theory.

There are numerous examples for the application of ANNs. With regards to the smart grid, they find application in terms of weather forecasting, since wind farms and PV power plants rely on weather alone for their power output (Maqsood et al., 2004; Ruppert et al., 2014). Load forecasting is also possible using ANNs (Liao and Tsao, 2006).

While ANNs can yield impressive results, incorporating them into a software requires carefully designed interfaces. Sculley et al. (2014) outline some of the factors that create technical debt in this regard. Chapter 4 will therefore outline the interfaces that avoid entanglement.

2.5 Boolean Algebra

The Boolean Algebra, which is named after George Boole, whose logical calculus created it in 1847 (Boole, 1847), is a special algebraic structure. It is, according to Peano (1888), a set with two elements, namely *true* (1) and *false* (0), on which a unary operation, the negation, *not* ($\neg$), and the two binary operations *and* ($\wedge$), and *or* ($\vee$) are defined. Other operations, such as the exclusive-or, *XOR* ($\oplus$), are defined using these operators and Peano's axioms.[55]

A binary vector of length n is a n-tuple

$$\boldsymbol{b} = (b_1, b_2, \ldots, b_k, \ldots, b_n) \text{ with } b_k \in \mathbb{B} = \{0, 1\} \ . \tag{2.11}$$

The Boolean space $\mathbb{B}^n$ contains all vectors $\boldsymbol{b}$ of length n:[56]

$$\mathbb{B}^n = \{\boldsymbol{b} \mid \boldsymbol{b} = (b_1, b_2, \ldots, b_k, \ldots, b_n) \text{ with } b_k \in \mathbb{B}\} \ . \tag{2.12}$$

This basic definition is important in order to define a Boolean algebra as an algebra of sets.[57] However, this definition alone does not yet enable efficient representation and solving of large Boolean equations. An equation $\mathrm{f}(\boldsymbol{x}) = \mathrm{g}(\boldsymbol{x})$ with n distinct variables requires us to compute 2^n Boolean values in order to achieve a complete solution set. Not only does this exhibit a runtime complexity of $\mathcal{O}(2^n)$, it can also, in the worst case, entail $n \cdot 2^n$ symbols in 2^n vectors that need to be stored.

Ternary Vector Lists (TVL) are a structure to efficiently store a list of *Binary Vectors* (BV). The space efficiency of a TVL is achieved through the

---

[55]More specifically, every Boolean Algebra gives rise to a ring through $a \oplus b := (a \wedge \neg b) \vee (b \wedge \neg a) = (a \vee b) \wedge \neg(a \wedge b)$.

[56]The number of elements in a given space $\mathbb{B}^n$ is therefore 2^n.

[57]In fact, Peano (1888) introduced the symbols $\cup$ and $\cap$; Huntington (1904, 1933b,a) introduced Boolean algebra as axiomatic algebraic structure in a series of papers.

introduction of a third symbol, '−.' The binary tuple $\mathbb{B} = \{0, 1\}$ is thus extended and becomes $\mathbb{T} = \{0, 1, -\}$. The two-valued nature of the original model is not lost; a biunique mapping between a *Binary Vector List* (BVL) and a TVL is ensured by the following rules:[58]

- The symbol '−' may be substituted by 0 as well as by 1.

- Every substitution of the dashes by 0 or 1 must lead to a binary vector of the original BVL.

- Each BV of the given BVL must be reconstructible from at least one *Ternary Vector* (TV) of the resulting TVL through an appropriate setting of the dash elements.

A researcher cannot only use TVLs as space-efficient data structure; operations with TVLs that are used to represent Boolean functions are also well-defined: At no point is it necessary to convert a TVL to its BVL representation when manipulating Boolean functions that are represented by a TVL. The standard monographs by Posthoff and Steinbach (1979b,a); Steinbach (1984) detail and discuss the usage of TVLs for that purpose.

The most important tool that implements TVLs and operations on them is XBOOLE (Bochmann and Steinbach, 1991; Steinbach, 1992; Dresig, 1992; Posthoff and Steinbach, 2004). The symbols used in this thesis, along with their corresponding XBOOLE function, are listed in Table 2.2.

XBOOLE has been used successfully to solve complex equation systems resulting from various models, including demand and supply of active power in the smart grid (Steinbach and Posthoff, 2012, 2014; Posthoff and Steinbach, 2014; Steinbach and Werner, 2014; Veith and Steinbach, 2015).

Another approach to handling complex Boolean functions are *Binary Decision Diagrams* (BDD). This Divide-et-Impera method is based on articles by Akers (1978); Bryant (1986). A BDD is a tree—i.e., a rooted, acyclic graph—that represents a Boolean function: Its nodes represent the function or its subfunctions and are labeled by its variables. Each node has exactly two edges: One that corresponds to the variable being set to *true*, i.e., the 1-edge, and another one a solver travels when the variable denoted by the vertex is *false*, i.e., the 0-edge.[59] The tree's leaves, or terminal nodes, represent the function's values. If edges from two distinct non-terminal nodes that share the same result

---

[58]Translated from Bochmann and Steinbach (1991)
[59]Bryant (1986) denotes the children of a vertex v as high(v) for v and low(v) for $\neg v$.

Table 2.2: Boolean operators, their set operator counterparts, and the corresponding XBOOLE function using Ternary Vector Lists in Orthogonal Disjunctive/Antivalent form

Name	Logic Notation	Set Notation	XBOOLE Function
Not	$h = \neg f$	$H = \bar{F}$	`H := CPL(F)`
And	$h = f \wedge g$	$H = F \cap G$	`H := ISC(F, G)`
Or	$h = f \vee g$	$H = F \cup G$	`H := UNI(F, G)`
XOR	$h = f \oplus g$	$H = F \triangle G$	`H := SYD(F, G)`

also share the same terminal node, and this is true for all, i.e., the BDD has at most two terminal nodes, it is also called a reduced BDD.

In his paper, Bryant does not only prove that graph manipulation operations can be used to work with Boolean functions that are represented as a graph, but also shows that the ordering of the variables has a huge influence on the graph's layout and size. Fig. 2.11 shows two graphs that denote two functions that differ only in the ordering of their input arguments, yet the BDD first one has 8 edges, while the second one features 16 edges.[60]

Since the original article, many special forms of BDDs have emerged. For this thesis, two types deserve special notice: *Zero-Suppressed Binary Decision Diagrams* (ZBDD) and *Edge-Valued Multi-valued Decision Diagrams* (EVMDD).

ZBDDs introduce nodes only when the positive part is different from the constant 0. This allows the user to create compact graphs for functions that yield 0 as result for most argument values. ZBDDs have been introduced by Minato (1993). He later applied the ZBDD concept to the power grid in order to find a feeder configuration that minimizes line losses in changing demand/supply scenarios (Inoue et al., 2014).

A *Multi-valued Decision Diagram* (MDD) represents a function f($\boldsymbol{x}$) whose arguments $\boldsymbol{x} = (x_1, x_2, \ldots, x_n)$ are p-valued. Its results are values from $(0, \ldots, p - 1)$ as well and thus constitute the tree's p terminal nodes, i.e., one corresponding to each logic value (Kam et al., 1998). A MDD is therefore useful to represent functions in the integer domain. A MDD can be reduced in the same way a BDD can be.

An *Edge-Valued Binary Decision Diagram* (EVBDD) is a directed, acyclic graph whose edges are valued: in addition to the edge denoting 0, an edge can have any value, not just 1 as it does in a BDD. The tree's non-terminal nodes

---

[60]Cf. Bryant (1986, Fig. 2).

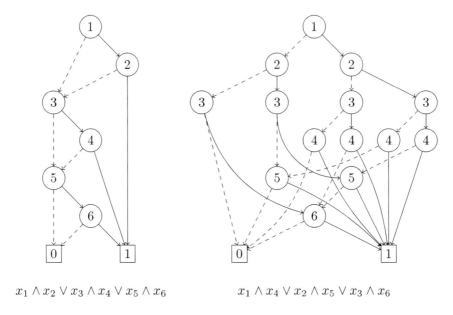

$$x_1 \wedge x_2 \vee x_3 \wedge x_4 \vee x_5 \wedge x_6 \qquad\qquad x_1 \wedge x_4 \vee x_2 \wedge x_5 \vee x_3 \wedge x_6$$

Figure 2.11: Binary Decision Diagrams of two equivalent Boolean functions with different argument ordering

denote the variables; its only terminal node is 0. The result is obtained by adding the edge values. EVBDDs are used for so-called *pseudo-Boolean functions*, i.e., functions that map Boolean arguments to integer results: $\{0,1\}^n \to \mathbb{Z}$. For functions with integer variables, these most be converted to BVs. EVBDDs were introduced by Lai (1993); Vrudhula et al. (1996).

EVMDDs, finally, shorten EVBDDs and MDDs by combining their strengths. An EVMDD can map an integer input to a multi-valued integer output: $\mathbb{Z} \to \mathbb{Z}$. This structure has been introduced by Nagayama and Sasao (2007) to represent elementary functions in a compact manner.

The fundamentals outlined in this section will find their application in the agent's heart piece, found in Chapter 7.

3 Approaching the Smart Grid by Modeling and Simulation

3.1 Models of the Power Grid

Traditional Power Plants

We know from Section 2.2 that a simulation is appropriate whenever the subject of research is too complex for calculation, or when a simulation will yield insight into the behavior or intrinsic nature of the subject in a new or altered environment. The power grid is such a subject: Not only is it complex, but the change in the power mix through the large-scale introduction of volatile renewable energy sources places it in an altered environment. More than that, the application of software agents as proposed by this thesis will alter not only power generation and consumption behavior, but also communication. Computer networks have been subject to simulation for a long time as has the power grid; this work will therefore follow this tradition.

A simulation that seeks to create an environment within which software agents control the distribution of power must also provide models of power generators and consumers.

One of the easier models that are available is that of a traditional power plant based on steam or gas turbines. Necessary details on the physical and engineering basics required for the modeling can be found in Section 2.1.

Such a traditional power plant is initially defined by its rated power output, which is also part of the standard vocabulary, e.g., 'a 1000 MW coal power plant.' In the power plant's type there are also, through its design and fuel, two other important parameters hidden: Its load gradient and the minimum load. The fourth parameter necessary for creating an initial model is the power plant's startup time, which typically depends on the time the plant has been offline.

Table 3.1: Parameters and values for traditional turbine-based power plants

Parameter		Bituminous Coal	Lignite	Gas	Nuclear
Rated Power (MW, per block)		500–1000	≤ 1000	≤ 340	600–1500
Load Gradient ($\%P_N$)		4	2.5	10–25	5
Minimum Load ($\%P_N$)		40	50	40	50
Startup Times	0–8 h	210	120	5–9	90
(min.,	8–48 h	225	240	10	210
by time offline)	> 48 h	300	600	10	1050

Values for these parameters, according to type, can be found in Table 3.1.[1]

In terms of software engineering, a separation of the model and its parameters from the concrete simulated subject's state is appropriate. The model is therefore defined by the parameter quadruple as listed by Table 3.1—the *startup times* member of the quadruplet being a triplet itself—, whereas the tuple (*Output, StateChange*) denotes its current state. This is depicted in Fig. 3.1 on Page 53.

Renewable Energy Sources

For the renewable energy sources, the parameter set differs depending on the actual source of power. A photovoltaic power station is characterized by its peak load and its relative output, subject to the solar radiation reaching the panels. Geothermal power plants are characterized by their rated output and are typically not subject to load control; they can thus be modeled as constant feeders in the simulated environment. Pump storage power plants are equally easy to model, given a rated power output, and their load gradient. No type of storage power plant can provide power indefinitely, for obvious reasons, and thus their storage capacity in terms of electric power constitutes the third necessary parameter. Since an electric power storage system can also act as consumer, the system's efficiency is of further importance, because it defines the conversion ratio between consumed power and potentially feedable power.

---

[1]The table is based on Brauner et al. (2012) and VDMA Power Systems (2013). Data about nuclear power plant rated outputs is based on data from the *International Atomic Energy Agency* (IAEA) PRIS database (International Atomic Energy Agency, 2015).

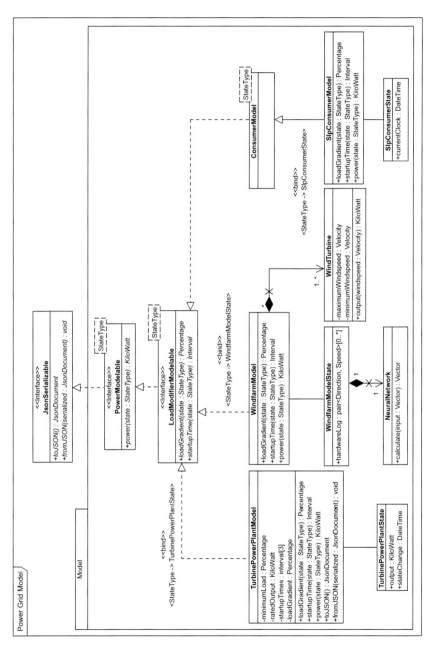

Figure 3.1: Excerpt from the simulator architecture for models and associated states

The models mentioned until now are relatively simple, given the model's parameters and the current state of the instance of the model. E.g., a coal power plant's power output, given a change request, is easily calculated from its current power output, its load gradient, and the minimum load boundary as seen in the previous section. Modeling a wind farm requires more complex calculations. Wind speed is not homogeneous in the area that a wind turbine's rotor sweeps; it obviously contributes the most to the overall output of the installation.[2] Additionally, wind speeds are typically measured at ten meters above ground. Also, the wind direction makes the model more complex considering a wind farm: If one wind turbine shadows another one, the second turbine's wind speed is the result of the fluid dynamics concerning the first one.

The alternative to a fluid dynamics simulation of a wind farm is to use an aggregated model (Slootweg et al., 2002; Pöller and Achilles, 2003), or to consider a wind farm a *black box* function, whose output depends on external parameters. This simulation environment uses the second approach. Here, an ANN is trained with data of an existing wind farm. Since ANNs can be trained to represent any function,[3] and the output of a wind farm can be considered as a function of each turbine's state, the wind speed and wind direction, an ANN can learn to represent a particular wind farm. RNN designs that incorporate the concept of time, such as Elman networks, can additionally learn to incorporate the wind farm's internal state, making this parameter unnecessary, leaving only wind speed and wind direction. Thus, we can describe the output P of the reference as:

$$P = \mathrm{f}(\boldsymbol{v}, \boldsymbol{\theta}) \, , \tag{3.1}$$

where the functional part is the ANN, and $\boldsymbol{v}$ and $\boldsymbol{\theta}$ are time series vectors[4] of wind speed and direction respectively. Note that, due to the time series effect, the ANN will need to be 'aware' of the concept of time, and thus an Elman RNN will be used. Here, the advantage of this structure as described in Section 2.4 becomes obvious: Wind turbines can follow the wind for a number of degrees, often slightly more than 720°. Afterwards, a motor must be used to rotate the nacelle back to its 0° position.[5] A perceptron could learn a time series pattern such as $((710°, 3400\,\mathrm{kW}), (720°, 3400\,\mathrm{kW}), 730°, 0\,\mathrm{kW}))$, but it will be confused by the following $(720°, 0\,\mathrm{kW})$ tuple. Additionally, the series

---

[2]Cf. Section 2.1.
[3]Cf. Section 2.4.
[4]$\boldsymbol{v} = (v_t, v_{t-1}, \ldots, v_{t-n}), \boldsymbol{\theta} = (\theta_t, \theta_{t-1}, \ldots, \theta_{t-n})$
[5]Cf. Section 2.1.

$\boldsymbol{\theta} = (710°, 720°, 710°)$ will constitute a new pattern for the perceptron, whereas the Elman RNN possesses a notion of the state context through its context layer.

In order to use this model to represent any wind farm, the ANN's output must be modified. If n is the number of turbines in the reference wind farm, n' the number of turbines of the modeled one, and the other wind farm consists of the same turbines, then the wind farm's output is constituted as

$$P = \frac{n'}{n} \mathrm{f}(\boldsymbol{v}, \boldsymbol{\theta}) \ . \tag{3.2}$$

This assumes that any modification of the reference wind farm would place the wind turbines in an optimal layout.

For each wind turbine that differs from the reference type, their power curve must be taken into account. If $\hat{\mathrm{p}}_k(v)$ is a function that maps a certain wind speed to the k-th turbine's output, and $\mathrm{p}(v)$ represents the reference turbines' power curve, we can modify the original Eq. (3.2) to:

$$P = \mathrm{f}(\boldsymbol{v}, \boldsymbol{\theta}) \sum_{k=1}^{n'} \frac{1}{n} \frac{\hat{\mathrm{p}}_k(v)}{\mathrm{p}(v)} \ . \tag{3.3}$$

Fig. 3.1 on Page 53 includes the UML notation for the class that represents a wind farm and its state. Here, the `WindfarmModel` class is associated with a number of objects instanciated from the `WindfarmModelState`, which, in turn, contain one RNN. Note that the RNN that represents the wind farm's model is part of the state class: This is due to the RNN's context layer that represents the wind farm's current state.

Power Grid Infrastructure

Models of the grid's infrastructure such as transformers or the lines themselves depend on their capacity, which may not be exceeded. The details have been discussed in Section 2.1. Thus, power lines and transformers are initially modelled by their impedance, voltage level, and maximum load capacity; the item's current load defines its state.

The impedances of lines and transformers can simply be given, but stem from different formulae. The transmission line's impedance is dependent on its length[6] and a number of parameters, which are:

---

[6] As long as the same unit of length is consistently used, the length's unit is not considered specifically, but only the length's value. We assume kilometers in this thesis.

R' Resistance per unit length per phase:[7] Ω/km

X' Reactance per unit length per phase: Ω/km

B' Shunt susceptance per unit length per phase: siemens/km

G' Shunt conductance per unit length per phase: siemens/km.

As soon as the agent software begins to work and short-term contracts on consumption and delivery of active or reactive power are struck, the simulated power grid diverges from a known state it initially had. Thus, a power flow analysis must be conducted whenever power generation and consumption changes.

The mathematical statement of the problem is extensively outlined and discussed by Powell (2005, Chapter 2); this thesis will now only summarize the statement in order to introduce the relevant solver algorithm.

Every node[8] i is either a generator or a consumer. A generator supplies active power and voltage to the system and is therefore called a *PV bus*.[9] At a consumer or load bus, active and reactive power values are known; it is therefore called a *PQ bus*. For the analysis, one bus is specially picked at which the voltage and the voltage phase angle are known; this bus is the *VD bus*.[10]

For every node i, we must consider the flow of current to or from its adjacent nodes:

$$\underline{I}_i = \sum_{k=1}^{n} \underline{I}_{ik} \; . \tag{3.4}$$

The flow of current is defined by the voltage difference between i and its k-th neighbor as well as the admittance of the grid elements between the two nodes. With $\underline{I}_{ik} = (\underline{V}_i - \underline{V}_k)\underline{Y}_{ik}$ we can rewrite Eq. (3.4) to:

$$\underline{I}_i = \sum_{k=1}^{n} (\underline{V}_i - \underline{V}_k)\underline{Y}_{ik} \; . \tag{3.5}$$

---

[7]We can represent a three-phase circuit by an equivalent single-phase circuit, if it is balanced.

[8]In the terminology of power system load flow analysis, nodes are called *buses* or *busbars*. We will continue to use the term *node* to maintain the information technology-centric view on the topic.

[9]Here, V actually denotes the voltage magnitude, $|\underline{V}|$.

[10]The VD bus is also known as the *slack bus*.

Note that all grid elements such as lines and transformers are descried by their admittance.[11] For lines and cables, any $\underline{Y}_{ik}$ is expressed by their resistance and reactance values:[12]

$$-\underline{Y}_{ik} = \frac{-R + \mathrm{j}X_L}{R^2 + X_L^2} \; .$$ (3.6)

A transformer is simply described by a simple series per-unit impedance.[13] By reformulating[14] the equations we can write them in matrix form,

$$\mathbf{I} = \mathbf{Y} \cdot \mathbf{V} \; ,$$ (3.7)

which forms the fundamental equation for each power system load flow analysis. The matrix notation introduces simplicity and also efficiency when creating a solver.

Voltage magnitudes and voltage phase angles of the nodes are interdependent, as Eq. (3.4) shows, and the power system load flow analysis is therefore non-linear in nature. Thus, the solver must employ numerical methods to arrive at a solution step-by-step. The simulation environment of this chapter uses the Newton-Raphson approach (Powell, 2005, Chapters 7 and 9) with the additional optimization factor described by Wang et al. (2014) to solving.

Power Consumers

Power consumers are modeled by SLPs, which have been described in Section 2.1. If large consumers can change the amount of power drawn, a load gradient is supplied for the model; it is then used as an offset to the original load profile. Additional modeling logic, such as shift schedules, can be included in derived model classes in order to create very specific simulations. All parameters for the models used in the simulation are summarized in Table 3.2.

The implementation separates the model and its parameters from the state of the model instance. This way, a model instance can be shared by different entities at different positions, where the corresponding simulation entity objects maintain the specific state. The states are serialized for evaluation of a simulation run and its results. Furthermore, replays are possible and model parameters

---

[11]$Y = Z^{-1}$.

[12]NB. that R' is the resistance per unit length, whereas R denotes the total resistance over the whole length of a cable, i.e., $R = R'l$. C.f. Section 2.1.

[13]As long as the wiring is fixed during the analysis.

[14]Cf. Powell (2005, p. 17).

Table 3.2: Parameters for different power grid models

Model	Parameters
Steam power plant	Startup times, rated power output, load gradient, minimum load
Photovoltaic power station	Peak power output, relative power output given solar radiation intensity and angle of incidence (as function)
Wind farm	Number of turbines, rated power of turbines, power curves of turbines (as function)
Power storage Consumer (modeled after SLP)	Rated power output, load gradient, efficiency Load profile (as function)

can be tweaked in order to find solutions to specific situations. For example, given a state, one can modify a model's parameter and answer questions like, 'how would the situation develop if this power plant was more flexible?'

Fig. 3.1 presents an excerpt of the simulator architecture implementation regarding models and states. All objects participating in the simulation are instantiated from classes that implement the `PowerModelable` interface as a uniform *Application Programing Interface* (API) that returns the current power balance for an object given its corresponding state. All objects the influence grid's power balance as a load—both, positive, i.e., as consumers, or negative, i.e., as generators—and that therefore feature a startup time if offline and a load gradient implement the `LoadModifierModelable`. Concrete classes and their states are then the `TurbinePowerPlantModel` for all traditional, turbine-based power plants such as coal, gas, or nuclear power plants, the `WindfarmModel`, and the SLP consumer.

3.2   Reference Situation

Assembling the models presented in the previous section, one can construct a reference situation. It becomes the starting point that illustrates the effects of volatile power generation while allowing the software agent and its modules to work in order to improve the overall power grid state. Illustrations of the agent's design and the individual modules in the following chapters will reference the

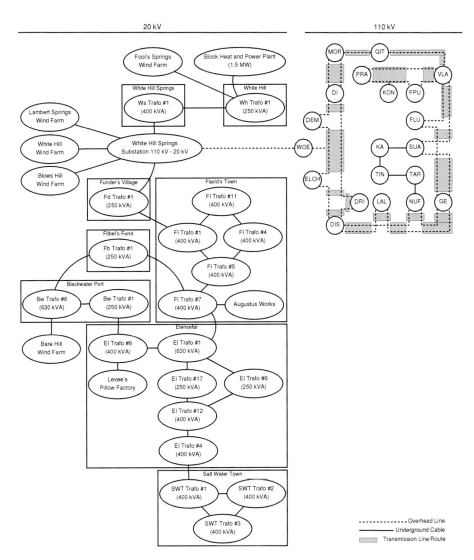

Figure 3.2: Logic view on the modeled reference power grid

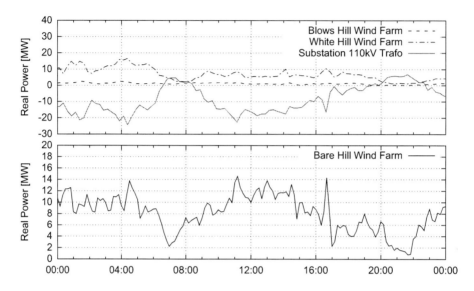

Figure 3.3: Active power at wind farms and the connecting substation during a day in February in the reference grid

situation presented here.

The reference power grid is based on an existing one in rural Germany, anonymized in this thesis.[15] The grid in question is operated at 20 kV, fed from the 110 kV grid layer. The substation's transformer connecting the two power levels can be loaded with at most 50 MVA. The smaller power stations in the grid range between 250 kVA to 630 kVA.

The 20 kV grid connects a number of villages as well as two small factories. Three wind farms are directly connected to the substation, another two are part of the grid near one of the villages, and therefore does not connect directly to the substation, but to another point in the 20 kV system. A small block heat and power plant completes the rural grid. Fig. 3.2 provides a logic view on it. Notice that in this figure a string of power substations has been merged into one if the consumers connected to it can be treated in an uniform manner and do not exceed the maximum load of the transformer.

---

[15]For easy reference, the villages, factories, etc. have been given identifiers in the form of new names that are easy to remember.

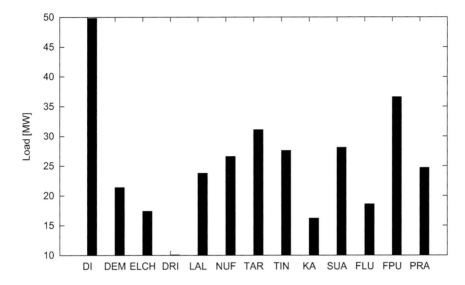

Figure 3.4: Peak load of the 110 kV nodes in the reference grid

The wind farms feed power whenever wind is available; there is no grid control. Equally, the factories do not participate in any form of control executed by the grid operator. The immediately obvious result of this policy is a surplus of active power that is available on windy days. Since no load adjustment of any form is done, the power is backfed via the substation's transformer to the 110 kV power grid.

The second diagram in Fig. 3.3 shows the output of the 'Bare Hill Wind Farm' during a day in February. The volatile output is typical for a wind farm in the area chosen as the reference area. A similar behavior is exhibited by the other two wind farms that are plotted, 'Blows Hill Wind Farm' and 'White Hill Wind Farm,' although neither of these is an immediate neighbor of the other.

Although the 'Bare Hill Wind Farm,' as shown in Fig. 3.3, is not directly connected to the substation but four villages away from it, the general shape of the power flow curve is recognizable on the active power flow curve of substation's transformer, which is also plotted in Fig. 3.3. The effect is more visible for the active power curve of 'White Hill Wind Farm' that is plotted in the same diagram. Clearly, the 20 kV to 110 kV transformer is feeding back.

The higher grid layer at 110 kV follows a ring structure. Each node in the grid has other 20 kV grids connected to it; they can serve to model sub-grids with different characteristics than the one displayed. The nodes 'MOR,' 'DIS,' and 'GE' can also serve as connections to a 220 kV or 380 kV grid to model large power plants. Most of the nodes in the 110 kV grid are connected using overhead lines of, on average, a length of 10 km.

Fig. 3.4 displays the typical peak load during a heavy-load period for the nodes of the 110 kV grid.

3.3   Smart Grid Simulation Environment

Motivation: Requirements to a Smart Grid Simulator

Any software needs to be tested before it is released to the general public. Testing provides the stakeholders with information regarding the quality of the software in concern. It helps not only to ensure the usability of the finished software in terms of human-computer interaction, but also to ascertain its correct behavior. This is even more important for software that is used in critical areas, meaning areas in which a malfunction of the employed software risks important infrastructure, people's lives, or both.

The electrical grid obviously is such an area. A loss of electrical power impedes nearly every kind of business; studies have shown that a civilization begins to show severe signs of degeneration after two days without electricity (Kutter and Rauner, 2012). Electricity does not only provide lighting, but is also essential to deliver fresh water and transport waste. Furthermore, it is needed for refrigeration and therefore required to supply cities with food. Malfunctions in the electricity grid can also harm people in direct, acute ways, for example, through excess voltages or effects of failing infrastructure.

The testing of a piece of software begins at the lowest level, where discrete units can be distinguished from each other. These *unit tests* isolate the parts of the program, testing them independently from the rest. A unit test is often a white-box test, meaning that the writer of the test code is also proficient with the code that is being tested, or at least has access to the sources.

A collection of unit tests that covers as much of the software as possible has several advantages. It ensures that a contract proposed by a given interface is fulfilled. Unit tests also show quickly when a change in one part of the software has led to a malfunction in another part, and thus facilitates change (e.g., via the refactoring of existing code).

Due to their isolated nature, unit tests are often executed using mock data. Mock data constitutes a set of input data that is specifically constructed in correspondence with the unit test. It serves to trigger specific behavior in the module that is subject to the test. Mock data can originate from real world data, but is often edited and tailored for the specific test case. However, mock data can also be totally synthetic.

While unit tests are focused on one particular, distinct part of the software, component and integration tests examine the cooperation of the different parts, up to considering the piece of software as a whole. All these test types serve to provide a deterministic, repeatable examination of the quality of the software. However, they operate on two premises:

1. A number of tests exist that describe the piece of software sufficiently to serve as a quasi-model of said piece of software with regards to the typical tasks.

2. The provided, mostly synthetic mock data covers enough variations of possible real world input data the software possibly encounters in order to trigger all important behaviors it will later exhibit.

Especially in cases where the first premise is not satisfied, a simulation needs to be employed additionally to any unit tests that exist. Recall Section 2.2, where additional motivation for employing a simulation is discussed.

The formal verification of any software is still an open question. The behavior of agents, especially distributed agent software that interacts with each other, is even harder to verify, although attempts at a sufficient notation and reasoning have been made.[16] The question of whether a model of the software and its behavior exists can thus be denied, requiring the additional testing a simulation can provide.

Testing an algorithm that controls a part of the electrical grid means a vastly changing amount of data is available, ranging from wind speeds and solar radiation to customer behavior. All these directly influence the agents, mostly through their forecasting; the combination of weather data, state of the grid, and consumer behavior cannot be represented in terms of mock data for one particular test case. The availability of real-world data and the complexity that arises from the interaction of the agents with each other, based on this data and their current state, mandates simulation runs.

Running a simulation instead of exposing the software to a real environment always alters the behavior of the software since the models that form the simulated environment are necessarily only as complex as required. Recall from

---

[16]Cf. Section 2.4.

Section 2.2 that model complexity, model confidence, and even model accuracy can stand in an anti-proportional relationship. But this does not only mean that some hitherto unknown relationship between these entities exists in the real world that is naturally missing from the simulated environment, but also that any simulation process alters the model's behavior through its simulated environment.

However, the creator of the simulation software should always strive to keep the influence of his simulation on the software as low as possible. Thus, a simulation should be conducted as black-box testing, meaning that the software that is subject to the tests should be treated as unknown, and the absence of errors should only be asserted by observing the software's behavior. The software that is subject to the simulation should need to be modified as little as possible, or not modified at best. Simply put, the development of the smart grid software agent must occur independently from the simulation software. This constitutes the first requirement of the simulation environment.

The second demand on the simulation environment is its ability to provide input data to the numerous items that are being simulated. This does not simply translate as fast access to a sufficiently great data store. Weather conditions as well as customer behavior are strongly dependent not only on the time of day, but also on their respective location.[17] Input data must therefore reach only those instances of the software for which it is relevant. Changing conditions there must have their *area of effect* properly represented in the simulation environment.

The input data will, in most cases, stem from very different sources: Different sources for, e.g., weather data exist with different sample sizes. Their measurements can also differ in terms of accuracy, or the area for which a certain measurement is valid. The same holds true for customer behavior or historic data of power generation. In general, a simulation of the electrical grid will have to cope with different data sources that offer data of potentially differing *data quality*.[18] Data quality is typically assessed by the researcher, and its influence on the result of a simulation run is quantified later. A simulation environment can assist in that respect by comparing the data sources registered for a simulation run, giving an estimate on the validity of the simulation's

---

[17]Cf. Section 2.1 and Section 2.1.

[18]Data quality is an abstract term for which several definitions exist; a fitting definition especially regarding simulation runs is "[Data quality is] [t]he state of completeness, validity, consistency, timeliness and accuracy that makes data appropriate for a specific use." (Schultze-Melling, 2010, p. 256)

results. Semi-automatic assessment of the input data quality, and its influence on the final results is the third requirement to the simulation software.

Since a simulator itself is a piece of software, it might seem obvious to model the scenarios that are being simulated in the software, too. However, this confuses the material with the tool: The description of a scenario, i.e., its parameters—the layout of the grid that is being simulated, its participants, start time, possible end times, etc.—depend on the software interface of the simulator itself. The concrete simulation itself, however, is not the concern of the simulation software; only its execution is. A description of the simulators initial state, its subjects, their layout, the general parameters governing the initial state, is therefore the fourth requirement.

The existence of a separate description of a simulation can also yield another useful feature: It additionally allows to formulate the expected outcome of a simulation run.

Architecture of the Simulation Environment

The agent described in Chapter 4, starting on Page 81, is a distributed software; therefore, it thus achieves its effect by the communication between the different instances. Since the usage of computer networks forms an essential part of the functioning of the agent, using simulation software suggests itself for communication networks.

In fact, Bush (2014) suggests that there are equivalent concepts in power flow as well as in packet flow; simulating one can help to understand the other. His book further emphasizes how increased integration of communication networks through sensors aid in monitoring the network state, enabling a detailed and more precise analysis of the electrical grid's health. While he does not suggest that traditional safety measures such as reclosers can be replaced by computer-operated and networked ensembles of sensors and switches, they could redefine the role of reclosers, fuses, and others as a secondary, backup role.

The simulation of communication networks helps to observe the behavior of a network protocol before releasing it to the general public and is often a viable alternative to setting up large testbeds with real hardware. Network simulators facilitate the set-up of a virtual testbed, the deployment of nodes implementing a certain protocol, pushing changes to the nodes and monitoring the flow of packets. A simulator for smart grid agent software needs to fulfill the requirements outlined in Section 3.3:

1. It should provide an interface to the simulated agent software in order to entangle the simulated object and the simulation as little as possible.

2. The simulator should provide an interface to read data from an external data store in order to provide the simulation environment with real-world weather data, customer behavior, etc. Using a database to write simulation logs and results to is additionally desirable.

3. It needs to support the geographical placement of a simulated object in order to define an area of effect for events that occur in the simulation environment.

4. All data injected into the simulation environment should not be trusted per se. The simulator should help the researcher to assert the quality of input data utilized during a simulation run in order to quantify the impact the different data have on the result of a simulation run.

The simulation kernel follows the design of OMNeT++'s kernel module.[19] Every simulation run is represented by a `Run` object that encapsulates the run's current state (i.e., pristine, set up, running, or finished) and aggregates all kernel components: The simulation run's description, the event loop and its associated event dispatchers, the list of subjects to the simulation, and the spatial index. These pieces are outlined in the following paragraphs and depicted in Fig. 3.5.

The event system, at its heart, consists of a queue of events, and a number of event dispatchers. The event queue is ordered not on a *First In, First Out* (FIFO)-basis, but sorted by the event's destined time of execution.

The simulation acts on an artificial, non-proportional time unit known as a tick. The current time/date in the simulation environment is determined by the event's scheduled time. In between events, nothing happens; the simulation can, theoretically, jump days without something happening in the simulated environment.[20] At the beginning of a tick, the event loop picks the first event from the queue, determines all objects it is to be delivered to, and dispatches it to these objects. The objects, in acting on the event, can generate events of their own; they are inserted into the event queue at the appropriate position designated by their time. If this time is the same as the current simulated time, those events are delivered too. The tick ends when all events at the given time are dispatched. The core event loop is described in pseudocode as Algorithm 1 on Page 68.

---

[19] Cf. Section 2.2.

[20] The simulation *clock*—cf. Section 2.2—is therefore monotonic, but not geometric.

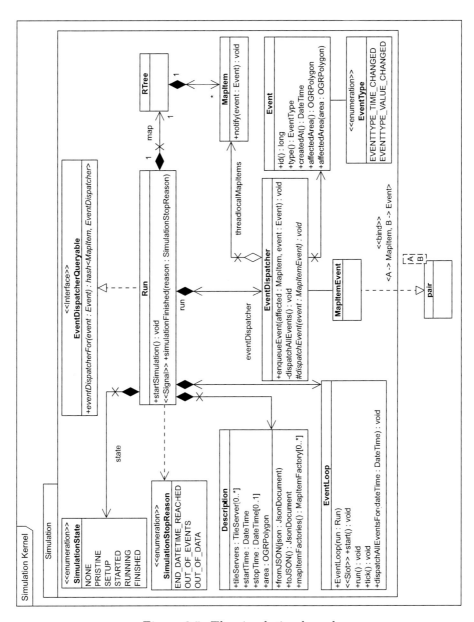

Figure 3.5: The simulation kernel

Algorithm 1 Event loop of the simulation software

---

procedure EVENTLOOP
    while $eventQueue \neq \emptyset \wedge eventQueue_{1,time} \leq stopTime$ **do**
        $e \leftarrow \text{POP}(eventQueue)$
        $currentSimTime \leftarrow e_{time}$
        $mapItems \leftarrow \text{GETALLIN}(e_{area})$
        for all $item \in mapItems$ **do**
            $dispatcher \leftarrow \text{EVENTDISPATCHERFOR}(item)$
            $\text{DISPATCH}(dispatcher, event, item)$
        end for
    end while
end procedure

The event dispatchers' raison d'être is to abstract the immediate delivery of events to objects; they can exist in terms of separate threads for a local, multi-threaded simulation, or serialize events and deliver them via a network connection to compute nodes. They handle thread boundaries or serialization transparently.

An event is delivered by calling the appropriate function on a `MapItem` object. Instances of this class serve two purposes: They act as wrapper classes to agent code and modelled grid infrastructure and therefore provide a unified interface to every simulated object. They also hold the position of the object in a geospatial sense, hence the name 'map item': It is about an item on a map.

The concept of the map item therefore fulfills two requirements: First, it provides a wrapper class with a uniform interface, allowing all simulated objects to be treated in the same manner, independently from the hardware they model, and, especially, independently from the code of the software agent, which is considered as a black box. Second, it allows events to carry their own geometry information in order to have an area of effect; the event's area is mapped to the map items' positions. Effective querying of the map for all items in a certain area is done via an R*-tree (Beckmann et al., 1990) geospatial index.[21]

Events can be created by any item on the map; not all `MapItem` objects also have tangible, real-world equivalents, such as power plants, or represent software agents. The map item concept is also used to inject data into the simulation. Thus, map items exist also to represent general customer behavior through

---

[21]Additional checking must be performed: An R*-tree organizes itself using rectangles, whereas any polygon can constitute the event's area of effect.

standard load profiles or weather conditions. These, too, can be geographically mapped, represent a modeled entity, and therefore fit the concept of the map item.

Since `MapItem` objects also serve as a way to inject data into the simulated environment, e.g., through wind speed measurements or solar radiation readings, they are also the common interface for any data source. There can be many types of data sources; but most will read their data from an external database and trigger events according to the measurements stored, or any kind of interpolated data stemming from interpolation points.

These data sources, although configured by the user, need additional scrutiny and may not be trusted necessarily. Data can come in different qualities, often because the initial collection of raw data yields sparse sets instead of continous series of high-quality readings. Even though these 'holes' in a particular data set exist, it is desirable to continue with the simulation instead of halting it: The behavior of the software agent observed during long-running simulations can lead to important findings; bugs sometimes lurk in code that is triggered only on edge conditions.

Additionally, different measurements describing the same subject can come from different sources. Such is the case with weather data that can be bought at high prices from national weather institutes, providing a high level of detail, in comparison to freely available weather data that provides only low-resolution[22] readings. If one was to plan a simulation running on real-world weather data spanning one year, one would either have to buy the whole year in terms of high-quality data, or resort to a full set of free, but potentially low-quality data. Instead, the simulator allows to mix both data sources in one map item. Due to an error assessment, the impact on the simulation result regarding the lower-quality data source can be quantified. Section 3.4, starting in Page 77, shows how this is done.

A simulation run needs to be configured before it can start. Configuring the simulation environment includes creating and setting up all map items, designating the start time of the simulation as well as a possible termination time, mapping the desired data sources, and, finally, potentially injecting custom nodes with pre-configured, artificial behavior that serve as decoys to trigger a specific reaction from other objects. This configuration[23] is done on

---

[22]E.g., wind speed and direction measurements are stored in terms of ten-minute mean values; this is the highest resolution (time-wise) one may obtain. Free weather data, however, often offers values averaged over one hour instead of every ten minutes.

[23]The actual run-time creation of objects is the task of Factory Pattern (Gamma et al., 1995b) classes that take their information from parts of the simulation description.

the basis of a *simulation description*, a JSON-serializable object that offers a separate description language that is not tied to the C++ code of the simulation environment. The `Description` object can also serve as the basis of a state comparison, wherein the final state of the simulation, or the final state of selected objects in the simulation, is compared to the desired state given in the description. This way, the results of a simulation run can be checked automatically against a given acceptance boundary.

The simulation description answers the requirement of separating of concerns. How the simulator uses it in order to conduct explicit testing is described in further detail in the following Section 3.3.

Abstract Simulation State Description

The simulation environment as it has been described until now is non-opinionated. The user sets the starting parameters—such as date and time or the objects participating in the simulation—, and, after starting the run, is able to observe the behavior of all nodes. Values are recorded for later review.

It is then up to the user to decide whether the simulation run yielded a success or not. He does so by comparing records of the time and taking error margins into account. The simulation environment itself is however completely unaware of the goals that are linked to a run and has no notion of a final state in terms of overall success or failure.

In order for the simulation environment to be able to assess this, it needs to be provided with a set of comparable properties with attached values. Therefore, a complete *description* of the simulation, its setup, and its final state, must be provided. This allows a uniform configuration of each simulation that enables the simulation environment to launch repeated runs from the same starting point. The behavior of the agent code can then be reliably tested and the impact of changes be determined.

Each description file uses the JSON as notation format, with one description being one object. Although the JSON is considered a dynamic format for which developers, in contrast to the XML, in practice do not provide a formal schema to verify JSON documents against, such a schema definition exists for the simulation description, created in *JSON Schema*.[24]

A description can be divided into three logical parts:

1. A general configuration section

---

[24]Draft version 4, cf. Zyp et al. (2013a) and Zyp et al. (2013b).

2. the initial state

3. the final state.

The *configuration* section serves as an information base in order to set up and configure the simulation run. It consists of several properties that are immediate children of the description object.

The user defines the *data sources* that are available during the simulation run via the `dataSources` property. Currently, two types of external sources can be configured: First, the project database that contains weather data, information about wind farms, and geospatial records containing the position of cities, power grid infrastructure and other items that can be contained in a simulation run. Second, an external *tile server* can be specified. A tile server makes pre-rendered tiles of a map available as images. This way, additional visual information like country borders or landscape features can be displayed.

What will actually be drawn from those defined sources is also configurable using the `fetchFromSources` property. This string list can contain any combination of the keywords `maptiles`, `consumers`, `producers`, and `grid`, whose presence enable the corresponding map feature from the defined data sources.

However, both the database and the tile server are optional data sources, because the description file itself can act as such. Using the simulation description file, a user can inject agents and connections between agents into the simulated environment. When no other data source is specified, this, together with the initial and final state definitions, can be used to set up synthetic black-box tests of the agent code. Whenever the user specifies the `agents` and `agentConnections` properties in the supplied description file, it is considered as a separate data source in addition to the external ones. Since the presence of these two properties obviously indicates the addition of injected items into the simulation environment, there is no formal need to explicitly specify the description file as a data source.

The artificially injected agents are identified by user-supplied strings that must be unique in the scope of the simulation run, but can otherwise be freely chosen. These ID strings are used wherever an agent as to be distinctively referenced, such as for the definition of inter-agent data connections using `agentConnections` property. Since ID strings are unique, but otherwise opaque as per the protocol definition, the agent's behavior is not influenced by this modification of their ID.

Whenever an agent is injected using the simulation description, the user can also supply its placement in the form of coordinates in WGS84[25] (Decker, 2000) notation. Otherwise, its placement is determined from the connections that are defined for this particular agent. When it becomes an item on the map, it needs a real position as outlined in Section 3.3. However, an artificially injected agent does not necessarily feature any meaningful position per se, and as such, it is generated using the spring-force algorithm outlined by Fruchterman and Reingold (1991) with the vertices being the data connections to other agents.

A simulation run starts at the earliest point in time for which data is available and continues until it has exhausted them. This seemingly simple notion is especially important in the case of injected agents with attached states, because it allows configured runs that test the functionality of a set of agents. However, in the face of a rich database, the user might want to select only a small time frame in order to, for example, test the agents' behavior under the influence of certain weather phenomena. For this, the simulation description can contain the optional `startTime` and `stopTime` parameters.[26] If no start time is given, the simulation starts at `1970-01-01T00:00:00Z`, i.e., the *epoch*.

In order to use the simulation environment for black-box testing, the user injects a partial initial state into the environment. A complete state of a simulation run is defined by the states of each agent participating in the simulation run:

$$S_t = \{A_{i,t} \mid i \in I\} \ . \tag{3.8}$$

Here, i is the identifier of each agent, t denotes the simulation clock.

The agents' states, in turn, are defined by the following quintuplet[27] denoted in Eq. (3.9) that includes the states of each agent's modules:

$$A_{i,t} = (F_{i,t}, P_{i,t}, M_{i,t}, C_{i,t}) \ . \tag{3.9}$$

Each module requires its own state definition. The *forecaster module's* state is defined by a set of tuples:

$$F_{i,t} = \{(t, P), \ldots\} \ . \tag{3.10}$$

---

[25] *World Geodetic System* (WGS)

[26] These parameters must follow the ISO 8601:2004 format (International Standards Organization (ISO), 2004).

[27] The initial definition can be found in Veith et al. (2014).

This denotes that the forecaster module assumes a power balance[28] of P at a time denoted by simulation clock t. The *power balance module's* state is defined analogously:

$$P_{i,t} = \{(t, P), \ldots\} . \tag{3.11}$$

Furthermore, the *messaging module's* state requires the notation of the relevant messages that reside in the agent's message journal.[29] The agent's message cache as well as the message format are described in Section 6.2; each message is noted literally:[30]

$$M_{i,t} = \{m_j \mid j \in J\} , \tag{3.12}$$

where J denotes the set of all message identifiers used throughout the simulation run.

Message IDs are generated by the sending agent; they are unique as well as opaque.[31] The latter property is used by the simulation state matcher to solve a conundrum: The scientist who writes the simulation run description wants to verify the correct behavior of the agents, and that will in most cases involve a check whether the desired messages were sent. However, each agent generates the message IDs independently from its environment, and in particular without knowledge about the simulation environment. Thus, it cannot by itself set the desired message ID that is used to verify the message's path. The simulation environment therefore records each message that travels through a data connection, matching it to all message definitions in the final states section of the simulation run description. Whenever a message matches, its ID is replaced by the one defined by the author of the simulation run description. This matching is done once for each unassociated message in the description's final states section; each match thus removes one unknown or unmatched message.

An actual message sent by an agent and a message description contained in the simulation run description are considered equal *if and only if* (iff) all fields defined in the description–with the exception of the ID iff the message description is part of the set of unmatched messages—match.

---

[28]NB. that the power value as hitherto defined means active power; the notation can easily extended to introduce reactive power by including the notion of a Q. Only for the sake of simplicity are reactive power values excluded from this description.

[29]The message journal of an agent contains all messages relevant to the agent's messaging functions at a given time. Cf. Section 6.2 and Section 6.2 for details with regards to the message journal.

[30]With regards to the semantics of the state matcher, see below.

[31]Cf. Section 6.2.

Finally, the notation of the *constraints module* is dependent on the actual constraint that the user needs to express; it must be a propositional logic term.

Since the purpose of all agents, and, in fact, the very definition of the term agent, relates to their ability to yield their own behavior, a simulation run does not necessarily need a complete state definition to execute.

Everything between the initial and the defined final state of the simulation is assumed to be calculable during a simulation run. The purpose of a so defined simulation is to ascertain success or failure, but not to provide a 'script' that is closely followed. As such, nodes that inject data such as weather conditions and are based on stochastic principles do not form an antithesis to this definition. It is necessary that all agents behave correctly in the face of a wide range of conditions. Since all generating nodes must be based on realistic principles, it can be assumed that all states generated by the combined data emission of all these nodes is also a realistic situation, even if it constitutes an aberration of weather, consumer, or other conditions.

Iff at least one final state is given, the simulation environment's state matcher is used to determine the success of the simulation run in terms of defined states. The state matcher ignores surplus information such as agents that are not defined in the description. If, however, a state definition exists, that of the simulation description is compared to that of the actual simulation run. Only an exact match constitutes a success of a particular check, and only if all matches executed are successful is the simulation run itself considered as being successful as far as the simulation state definition is concerned.

The notion of ignoring surplus information instead of forcing an exact match over all properties serves two purposes: First, the user can omit information that is not necessary for determining the success of the simulation. For example, during a test of a demand/supply calculation algorithm, the user might want to check whether certain offer notification messages have been received and answered, but not what the distance they have travelled in terms of a hop count is. The user can therefore design test cases as to its intent instead of its technical necessities. Second, it allows mixing agents defined in the simulation description file with those originating from another data source, such as a database, without having to define all agents from all data sources.

The state matcher can work on a set basis for all elements, in all modules, following the general pattern of

$$success = \begin{cases} true & \text{iff } S_D \subseteq S_T \text{ ,} \\ false & \text{otherwise.} \end{cases} \qquad (3.13)$$

This means that every agent's state, as defined in the simulation run description, needs to be sub-matched in a similar manner, i.e.,

$$success = \begin{cases} \text{true} & \text{if } \forall A_i \mid i \in I' : \forall X_{i,D} \subseteq X_{i,T} \mid X \in \{F, P, M, C\}, \\ \text{false} & \text{otherwise.} \end{cases} \quad (3.14)$$

This continues recursively until the atoms of the corresponding terms are compared for equality.

Using the simulation description, repeated runs of test cases in the simulation environment are possible without additional user setup. The user creates the simulation description once, but can use it to verify the process of his agent software code in simulation runs. Additionally, completely synthetic simulation set ups are also possible.

External Data Sources

The previous Section 3.3 showed how a simulation description does not only implement a separation of concerns, but also allows the configuration of different data sources.

Data sources allow the injection of any data, be it a sensor reading, wind speeds, solar radiation, or SLPs for customer behavior into the simulation environment. Section 3.3 described the `MapItem` class as the abstraction of all participants of a simulation run. Data sources are—from the organizational perspective—map items, too. A certain area for which the data of the source is valid, i.e., where it is effective, is the attribute all sources share. Thus, there is a piece of geometry information attached to each external source of data, which in turn marks it for representation as a map item in the simulation.

Fig. 3.6 shows different areas of effect for a number of data sources. The shaded parts of the image depict a data source and the polygon that represents the area for which the data it provides is valid. Notice that aside from the highlighted area, the data source objects are not represented by any icon, nor is there any other indication that these map items are in any way special.

Data sources act on events just like any other item in the simulation environment, too. Their primary trigger is the new simulation time, propagated through any `Event` that carries date/time information. Looking back at Fig. 3.5, the `EventType` enumeration lists the `EVENTTYPE_TIME_CHANGED` that carries exactly this information. The data source will, upon reception of such an event, look up the data associated with the given time and create an event of its own. Such events are of the `EVENTTYPE_VALUE_CHANGED` type.

Figure 3.6: Data sources and their area of effect

The value-changed event carries the area of the data source as its own area of effect. It reaches, through the *Geospatial Information System* (GIS) index lookup, all other map items that are within this area. How the receivers react depends on what they model: A change in the wind speed is interesting only to models of a wind farm, whereas changes in customer behavior through a standard load profile reach only models of villages, factories or other models of consumers. Filtering is done through simple tagging.

Whenever a data source receives its own value changed event, it looks up the next simulation clock at which a new event of the same type—but with a new value—must be scheduled. The actual lookup depends on the data source; a new value for wind speed and direction from a list of mean values will be interpolated linearly given a desired Δv of $1\,\text{m/s}$.

This circle is broken whenever a data source runs out of data. In this case, an exception of the type `DataSourceDepletedException` is thrown. A depleted data source does not immediately force the simulation to halt. Instead, the map item that represents the data source checks if an auxiliary repository is available.

In fact, data sources of different quality are stacked on each other. The source with the highest quality also has the highest priority assigned; the ordering of these repositories is done by the user. Once a data source is depleted and the exception is thrown, the simulator, through the map item, uses the next repository that is available in the stack.

This allows for the simulation run to continue even if the high-quality data source is no longer available, but auxiliary ones are. Continuing in spite of a depleted source allows to expose the agent code to a higher variety of situations while still maintaining the overall grid state that is the result of the current simulation run. While worthwhile, switching data sources has implications for the overall outcome of a simulation run. The next section will therefore describe how an assessment of data quality is made and how this assessment is reflected in the results.

3.4 Data Quality Assessment and its Influence on Simulation Runs

The Winzent simulation environment allows stacking of data sources of different quality, as detailed in the previous section. The immediately obvious advantage this provides is a continuation of a simulation run even if the high-quality data source is depleted. This is especially interesting in cases where data of higher quality is available, but must be bought—and may be expensive—, whereas data of lower quality is available for free. Typically, a scientist has to set up two different simulation runs in this case; with the data source stack, this is not necessary.

However, this transition may not happen without notice: It is highly probable that the lower-quality data source will distort the overall result of the simulation run compared with the result obtained if the high-quality data source was available for the whole run. Usually, the scientist must assert the impact.

Ultimately, it is therefore a question of confidence in the result. If the status of a data source as the one with the highest data quality has been asserted manually, i.e., by the scientist before the simulation run was started, a run using only this source has a confidence of 100 %.[32] Results obtained with other data sources will have a lower confidence, naturally.

This abstract notion of confidence stems from two ways two data sources can derivate from each other. Considering that both sources try to describe the same subject—for example, readings of wind speed and direction—, they influence an event-discrete simulation in the number of events that are introduced through

---

[32]Relative to whatever confidence the scientist, through his knowledge or experience, has of this data source

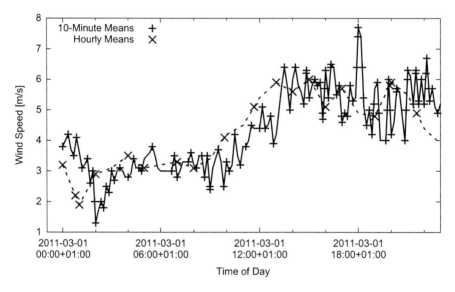

Figure 3.7: Events created by two different data sources for the same day

the data source and the derivation of values exhibited by the sources at the same simulation clock. Compare the two lines in Fig. 3.7: The number of events created by the first data source is significantly larger than the number of events created by the second one ($111 > 36$). Since obviously no action is performed in an event-discrete simulation environment when no events are scheduled, this observation directly impacts on the simulation run's result.

Assuming that both data sources have witnessed the same events and made the same measurements, but represent them in a different level of detail, we can understand them as two different encoders of the same source. Hence, the rate distortion theorem can be applied (Shannon, 1948, 1959).

The values observed by the two data sources are encoded as symbols of an alphabet. We assume that the high-quality data source encodes losslessly.[33] Therefore, the original symbol q of the alphabet Σ is compressed to the symbol $\hat{q}$:

---

[33]This is, of course, relative: If wind speeds are presented as ten-minute mean values by the high-quality data source, this encoding is, through the calculation of the mean value, not lossless. However, since this high-quality data source forms the reference, and we have no better source available, we assume this source to be lossless.

$$q \in \Sigma \ , \tag{3.15}$$

$$\hat{q} \in \hat{\Sigma} \ . \tag{3.16}$$

The channel itself is assumed to be lossless. For each pair $(q, \hat{q})$, we define the distortion to be:

$$\mathrm{d}(q, \hat{q}) = (q - \hat{q})^2 \ . \tag{3.17}$$

We can then compute the overall distortion of the data source $\hat{\varsigma}$ given the probability that q is emitted by the high-quality, reference data source ς, as $\mathrm{P}(q)$ and the probability that the distorted symbol $\hat{q}$ is emitted by the other data source instead as $\mathrm{P}(\hat{q}|q)$:

$$d_{\hat{\varsigma}} = \sum_{q} \sum_{\hat{q}} \mathrm{P}(q) \cdot \mathrm{P}(\hat{q}|q) \cdot \mathrm{d}(q, \hat{q}) \ . \tag{3.18}$$

This distortion can be computed for all data sources. The distortion values for all data sources $\hat{\varsigma}_1, \hat{\varsigma}_2, \ldots, \hat{\varsigma}_n$ can then be treated as error in terms of a mathematical error analysis. Thus, the result of each computation will additionally be augmented by its error. Since each power balance that matters to the user will be checked by the state matcher described in the previous Section 3.3, we can derive a total error from these balances:

$$\Delta P = \sum_{i \in I'} |\Delta P_i| \ . \tag{3.19}$$

4 The Universal Grid Agent

4.1 Modular Design Principle

The premise of this thesis is a mediation between volatile power generation and power consumption: Given enough information to forecast future demand and supply and consumers—and producers alike—that can accommodate a certain flexibility, we can create a stable equilibrium of the active and reactive power balance through a grid-wide planning process. However, in order to match information from forecasts and potential load gradients, massive capacities for information processing and computation are required.

Table 4.1[1] gives an estimation of all relevant actors in the German power grid. If one aggregates the low-voltage level on the transformers connected to the medium voltage grid,[2] assuming that the patterns that emerge from power generation or consumption can be aggregated at those transformers—making them, in effect, the most important node in a power grid from an information point of view—, an astonishing number of 561 069, i.e., more than half a million nodes exist in the German power grid alone. These nodes will log data: tuples of timestamps[3] and double-precision floating-point numbers[4] that will take up at least 128 bits each, if a simple tuple is enough to capture the condition of a node at a certain point in time. Usually, more than a tuple is required; for example, a wind farm will need not just a timestamp and a floating-point number, but several: Wind speed, wind direction, and real power output are necessary at least in order to produce a meaningful forecast, as Chapter 5 will show. While this might not seem much, an item's position must be saved, thus a GIS is

---

[1]Cf. Bundesnetzagentur (2015); dena — Deutsche Energie-Agentur (2013).
[2]Cf. Table 2.1.
[3]A 64 bit integer
[4]Also 64 bit wide

Table 4.1: German power grid infrastructure

Amount	Description
685	Wind Farms (onshore & offshore, ≥ 10 MW)
185	PV installations
272	Other renewable energy sources (e.g., bio-, geothermal energy)
372	Turbine Power Plants (coal, gas, nuclear)
559 555	Transformers (≥ 10 kV grid voltage)
561 069	Total

needed, taking up additional data space. Databases require additional data storage through the use of indexes to allow for efficient query execution. Also, the forecasting algorithms will need to store and serialize state information, like situation-specific trained ANNs.[5]

While storing and retrieving data might certainly be possible, potentially requiring a cluster of servers, the coordination of a large number of nodes with the goal to create said equilibrium of active power—or, reactive power, respectively—will need parallel processing power as well. This hypothetical compute cluster must also be well connected: Sensor data from nodes must constantly fill the data base in order to allow the forecasting algorithms to adapt to the current situation. In addition, redundancy must be built into the system architecture in order to avoid creating a single point of failure.

Instead of tackling the problem presented by a centralized approach that requires means of data storage and computing power appropriate to the task, another approach can be taken. Subdividing the problem into a number of smaller ones presents a valid alternative. This *divide-et-impera* solution distributes the information and processing load among all nodes in the power grid. In the example given above, the divide-et-impera approach will create half a million compute units of a small size. This will allow an operator to extend the system easily: Adding a node to the power grid implies adding the associated compute unit as well.

The approach solves the problem presented on an architectural level, but requires careful design of a communications architecture that allows the interchange of relevant information and computation results. In addition, a model

---

[5]Cf. Section 2.4 on ANNs in general and RNNs in particular and as to why each site will need to serialize an individual ANN.

of the environment of each node must be devised, since no shared knowledge exists automatically.

We can thus identify a number of tasks whose fulfillment is necessary for the proper functioning of the divide-et-impera approach. The architecture presented in this thesis models each distinct task as a module. Their entirety is shown in Fig. 4.1.

Modules are stacked in layers. These layers do not only represent the level of abstraction, they also express the priority of the modules. Modules situated on a lower level have precedence over those on higher levels. Thus, the software that immediately interfaces with the underlying hardware may override decisions of a high-level module if it endangers the machine, acting as a fail-safe design. The following modules make up the agent design:

Automatic Hardware Control This module is not part of the agent itself, but designates automation technology that is already present in the node. It is included to signify that the agent does not replace any existing part of the node's control mechanisms, but acts on top of it. This module has the highest priority, meaning that the agent must adapt to the idiosyncrasies of machine and place.

Hardware Interface This module interfaces the agent with the hardware. On a computer system, one would place the 'driver' at this point. The hardware interface is situated on the device layer of the module layout, as well as the logging module: These two will need to be modified in order to adapt the agent software to the node; all other modules use data from the lower ones.

Logging Retrieves data from the hardware and stores it in a log that is accessible to the module in the next upper layer.

Data Extraction The data extraction module uses data retrieved by the logging module to feed learner and forecaster. It thus serves as a filter and syndication module.

Learner The learner trains and re-trains the forecaster in order to tune it to changing situations, such as the current weather conditions.

Forecaster This module uses current sensory data, as syndicated by the data extraction module, to forecast demand and supply in the future. The temporary distance for which it can forecast is specific for each agent and dependant on the actual source of sensor data.

Constraint Calculation Constraints allow considerations from external sources that are not the result of the node it represents or its environment. Such constraints might include contracts where a certain supplier or customer is preferred, or a weak local line that is not represented by a dedicated agent. The constraints module is obviously an interface for external influence and will therefore experience no further consideration in this thesis.[6]

Reserve The reserve module represents allowable variance in terms of real or reactive power: It influences the power balance found in the Demand-Supply module, which normally tries to create an equilibrium of $\Delta P = 0$ and $\Delta Q = 0$.

Demand-Supply This is the keystone of the agent: It represents the power balance—real and reactive power—of the node the agent represents. Here, the agent stores forecasted deviations from the power equilibrium as well as requests and responses from other agents. It triggers the social behavior of the agent and solves the power balance when requests and offers from other nodes arrive.

Messaging The social aspect of the agent adheres to the protocol rules specified in Chapter 6 and takes care of the proper encoding. The messaging module feeds the Demand-Supply module with initial requests and responses that arrive to solve an imbalance.

Although some modules allow external exercise of influence, all agents can be treated uniformly: The messaging module represents each to the outside world. Every agent is uniquely identifiable through an opaque identifier. Chapter 6 will give more detail on options regarding the format of the identifier.

4.2   Interfaces

The layers, in addition to the level of abstraction and priority, also document the communication interfaces of each module. The agent design allows data flow only between modules in the same layer or between adjacent layers. Staying true to the layer concept, interfaces at lower layers deal with low-level data that is near to the actual hardware, whereas the information flow at higher layers clearly show abstraction.

---

[6]Cf. also Section 3.3, in which the constraints module is not further specified in its representation.

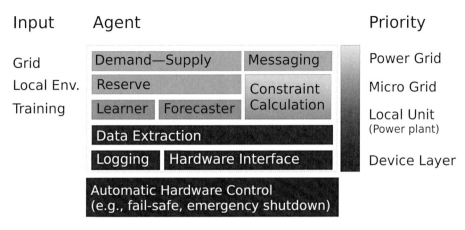

Figure 4.1: Modules of the Universal Smart Grid Agent

Consider Fig. 4.2. The hardware adapter uses the `HardwareBackend` interface that represents the driver and is unique for a certain type of node. Through `LogEntries` it communicates the current state of the node to the logger that provides facilities to store this data in a structured format—the `Journal`. This journal does not yet resemble a database; the data extractor offers the `Query` interface to upper layers.

The uniform `Query` interface allows the learner to train the forecaster—it uses a specific `TrainingAlgorithm` for this—as well as the latter to retrieve sensory information that forms the basis for the next forecast, which is amended by an `ErrorBoundary` from the constraints calculator.

A `Requirement` that encapsulates messages from remote agents, as well as forecasts from the local instance, is a piece of information the demand-supply solver requires to maintain the equilibrium of active power[7] stored in the `PowerBalance`. How much deviation from a mathematical equilibrium of $\Delta P = 0\,\mathrm{kW}$[8] is allowed, is expressed by a `Variance`.

The Governor interprets the `PowerBalance`, turning it into `Message` instances to express a demand for power or offers of surplus power. These messages are sent via the communication `Hub`, which implements the protocol rules.

---

[7]Or, reactive power, respectively
[8]Or $\Delta Q = 0\,\mathrm{kVAr}$, respectively

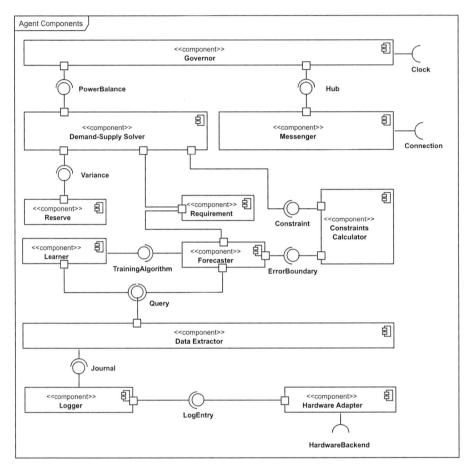

Figure 4.2: Components of the Universal Smart Grid Agent

The classes that form the implementation of the module concepts and their interfaces are detailed in the following chapters: Chapter 5 is concerned with the Learner and Forecaster. Chapter 6 details the communication architecture, i.e., the social part of the agent. Finally, Chapter 7 details the demand-supply solver.

4.3   Agent Behavior

An agent acts upon two incentives:

1. from the creation of a new forecast that indicates a disequilibrium of the local power balance

2. from the reception of a request originating from another agent.

Although it acts upon other message types, as detailed in Chapter 6, these two events are responsible for updating the agent's internal state—i.e., its model of the world—and form the agent's incentive to act. These two events signify a deviation from the equilibrium, the first one on the local node, the second on another node in the grid. The agent's primary goal is the conservation of the power balance, i.e., the equilibrium; thus, it must act if an event shows the arrival of the situation of a disequilibrium.

A forecast is generated at certain, even-spaced intervals. If the forecast indicates a deviation from a power balance, it triggers a request to other agents. Such a request can either indicate a demand for power—active or reactive—, or the availability of surplus power. In the latter case, the agent requests that the power is being used elsewhere: This will commonly happen when renewable energy sources cause oversupply.

Other agents receive the request and act upon it, ultimately sending offers to the requester. From these offers, the requesting agent tries to generate a solution that ensures the continuance of the power equilibrium. Only if no solution can be found may the agent exert local control, e.g., by throttling wind turbines or similar. Since this is a result that should be avoided—the agent may neither waste power nor produce a black- or brownout—, it is summarized under the term 'emergency measures' in the context of the Universal Agent.

Upon the second case, i.e., when a request is received, the agent must work to preserve the equilibrium at the remote node,[9] generating a forecast when possible within acceptable error bounds. If it can then send a response, it must do so.

A special case exists whenever a forecast—which is generated automatically when acceptable within defined error bounds—generates a request, which is met by another request, i.e., originating from another agent, which could actually be

---

[9]An agent cannot choose to not help a remote agent: Due to the nature of the power grid, a disequilibrium that goes ignored will eventually affect all agents negatively. This motivates the imperative: The agent must work to preserve the power equilibrium at any node.

a part of a solution. Therefore, two requests exist where neither is a response to the other one, but both would partially or even completely cancel each other. Then, one agent must withdraw its request and send a response instead. The algorithm underlying the resolution of this conflict is part of the protocol.

In order to function, the agent requires two models. One is the model of the node it represents, i.e., its local environment. This is the local power balance.[10] The other is the social model, i.e., that of its 'fellow' agents: This is partly represented in the power balance that gathers requests and offers for the finding of a solution to a potential disequilibrium, but also in the message journal[11] that stores messages received from remote agents in order to save their state for the amount of time required to arrive at a new equilibrium.

These models, along with the agent's overall behavior, are depicted in Fig. 4.3.

Note, that the agent's behavior cannot be described in terms of a *Finite State Machine* (FSM). An FSM's memory is represented by the set of its states, $S = \{s_1, s_2, \ldots, s_n\}$.[12] This is most obvious when an acceptor is modeled. Strictly speaking, the agent, whose goal is to calculate demand and supply and therefore has many different values for timestamps and power values, requires the notion of a general data store.

Abstracting this data store away might seem possible by modeling a message received as input and allowing states such as 'message saved in cache,' or 'forecast value saved in power balance.' Solving the power balance to an equilibrium by selecting the appropriate responses, however, entails that the requesting agent must contact the responding agents in order to notify them whether it indeed intends to form a contract with them. Thus, a memory that connects agents, their messages, and the offered values is required.

Furthermore, consider the following situation: An agent, A_1, offers a surplus of active power to the grid. According to the rules of the protocol,[13] this must be formulated as a request: A_1 requires that its surplus is consumed by another agent. Now, the message A_1 sends—denoted in this example by m_1—travels a certain time through the network, since no message can be transmitted via a real computer network without delay. Within the time frame A_1's message needs to find an agent that can present a demand to its offer, another agent, A_2, forecasts a demand for active power and sends a request of its own, m_2.

---

[10]Cf. Section 7.1.
[11]Cf. Section 6.2.
[12]Cf. Booth (1967) for an introduction to FSM.
[13]Cf. Chapter 6.

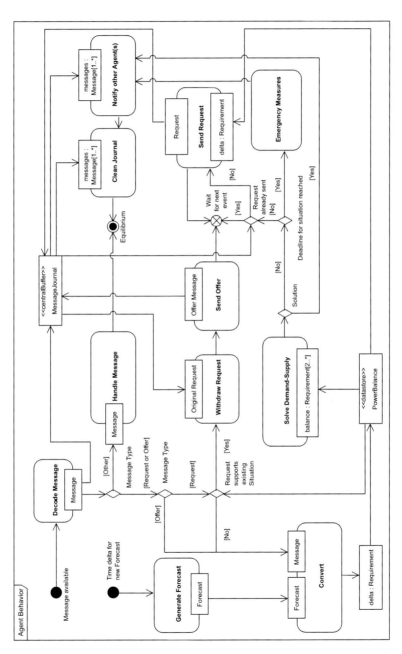

Figure 4.3: Activity diagram depicting the Universal Smart Grid Agent's behavior

Both requests m_1 and m_2 happen to be subject to the same time frame, i.e., m_2 would—possibly partly—solve A_1's request and vice versa.

In order to avoid a deadlock, one of the two agents needs to withdraw its request and answer with an offer to the other agent's request.[14] This requires two memory devices:

1. The power balance of both A_1 and A_2 in order to deduce that $m_{1,timespan}$ contains $m_{2,timespan}$ or vice versa

2. the message journal to map $m_2 \rightarrow A_2$ in order to formulate an answer

3. the message journal to withdraw m_1 with m_3, i.e., a memory to map $m_2 \rightarrow m_3$

4. the power balance to formulate an offer to m_2 for A_2: $m_4 \rightarrow m_2$.

These distinct actions require identification of agents and messages as well as the modeling of a certain situation that is only triggered by the coincidence of two requests with opposite ΔP[15] values. A modeled trigger for this situation is required that identifies its occurrence, which requires a memory that is

1. variable due to the individual timestamps, power values, and agents, but

2. precise in mapping agents and requests.

Thus, an FSM cannot be used to model the behavior of the Universal Agent; it requires a Turing machine that implements the algorithm depicted in Fig. 4.3, $q.e.d.$

---

[14]The details to the resolution to this conflict are described in the *Match-or-Forward Directive* in Section 6.2.

[15]Or, naturally, ΔQ

5 Forecasting Power Demand and Supply

5.1 Design of the Forecaster Universal Smart Grid Agent Module

Data Pipeline

Forecasting is vital to the work of the Universal Agent: Only if a divergence from the equilibrium is noticed before it occurs can a distributed planning process help to maintain the power balance in the grid. Fig. 5.1 on Page 94 therefore presents the forecaster module along with the necessary data structures. Obviously, the effort concentrates on the `forecast()` method that actually creates said forecast.

The agent, in its forecaster module, essentially needs to orchestrate a flow of data in order to extract and combine information as necessary to create a forecast that can be used in the context of maintaining the node's power equilibrium. For this, we can identify several key components:

1. The agent requires a data base that contains the raw data necessary to create a forecast.

2. Further, it needs an object that extracts the raw data and combines it, adding time values to it, in order to create a model of the node's state at a given time. Remember that creating a model of an agent's surroundings is an essential part of an agent's architecture.[1]

3. This data must then be fed to the ANN in use, requiring conversion from the correlated data to input and output vectors suitable for the ANN.

---

[1]Cf. Section 2.4.

The agent's representation of the node's state must be converted to the ANN's representation of it, and back again.

Initially, in order to create a forecast, a log of power balances at a given time are immediately necessary. If the power consumption or generation follows a time pattern, this is also the only necessary data required. However, additional pieces of information can enhance the accuracy of the forecast: E.g., data of solar radiation, wind speeds, or temperature might not be required to create a load forecast for domestic households, but since weather conditions influence people's power consumption, they are useful. In effect, they can help to detect abnormalities beforehand.

This forms the data base of the forecast. A usual set of information is useful and potentially extractable from the system's logs to form the data base includes, but is not limited to:[2]

1. Active power in kilowatts

2. reactive power in kilo-voltamperes

3. wind speed in meters per second

4. wind direction in degrees

5. solar radiation in watts per square meter

6. barometric pressure in millibar

7. temperature in Kelvin

In effect, the data extractor, whose purpose has already been summarized in the context of Section 4.1, forms a list of possible input data that can also contain additional information that is site-specific, for example, the orientation of a wind turbine's nacelle. The bootstrapping regarding which data is available and sensible can be performed in numerous ways, for example:

- Utilizing a sensor hardware discovery protocol (Hyyryläinen and Jantunen, 2006)

---

[2]NB. that currently, active and reactive power are the readings that should be available everywhere sensors are installed, whereas the rest, although it might be beneficial and give additional data to correlate when forecasting, is usually only available if already required for the operation of the hardware.

- querying neighbor sensors using the OSGP[3] or IEC 61850 (Brunner, 2008)

- using a fixed list of sensible data

- as a fallback, through manual configuration.[4]

The `DataExtractor` creates `JournalEntry` objects from the system log. These journal entries are sorted by a timespan, $\tilde{t} = [t_1; t_2)$,[5] for which the readings contained in the entry are valid. The `JournalEntry` objects are the immediate data objects in use in the pipeline and constitute the agent's model of its environment. The collection of relevant journal entries is stored in the agent's `Journal`.

The Universal Agent proposed in this thesis uses an unit system in order to define allowed operations and provide type safety.[6]

The components mentioned hitherto and an excerpt of the unit system are depicted in Fig. 5.1. Note that the part of the unit system depicted is that required for the reference situation presented in Section 3.2.

The journal entries can now not only be used to reason about the agent's current or previous state, but also to create the actual forecast. However, the RNN can not offer an immediate interface that takes a `JournalEntry` for input. This is not simply a question of interface design, but rather stems from the different modes of representation the agent through its journal and the RNN have.

The agent's journal maps a timespan to a number of sensor readings. Every journal entry is unique, i.e., at a given time there is exactly one reading for each sensor. Additionally, these sensor readings are absolute values. These two properties make it at first incompatible with the RNN[7] that operates with linear and tanh functions:

---

[3]Cf. Section 2.3.

[4]Node-local finetuning can be part of a service contract.

[5]Given $\tilde{t}_1 = [t_1; t_2)$ and $\tilde{t}_2 = [t_3; t_4)$,

$$\tilde{t}_1 \leq \tilde{t}_2 \quad \Leftrightarrow \quad t_1 \leq t_3 \ .$$

[6]Failure to provide an unit system—i.e., relying on the built-in types float, integer, etc.—has been proven to lead to catastrophic failures, such as the disintegration of the Mars Climate Orbiter, where a floating-point number was interpreted as representing miles in one function and kilometers in another (Euler, 2001).

[7]Applicable to ANNs in general

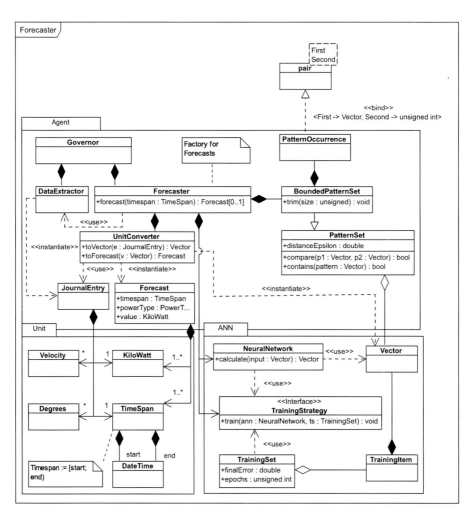

Figure 5.1: Forecaster module of the Universal Smart Grid Agent

$$f(x) = x, \qquad\qquad -1 \leq x \leq 1 \,, \qquad (5.1)$$

$$\tanh(x) = 1 - \frac{2}{e^{2x} + 1}, \qquad\qquad -1 \leq x \leq 1 \,. \qquad (5.2)$$

Eq. (5.2) with its limited domain is able to represent the input interval $(-100\,\%; +100\,\%)$. The agent must therefore convert absolute values from sensor data to relative values, in both directions: First to create the input of the RNN and afterwards to retrieve the result. This conversion is handled by a number of `UnitConverter` objects that bijectively convert a `JournalEntry` object into a vector of floating-point numbers that form the input and output of said RNN.

For each distinct sensor reading that is part of the entry, a corresponding input layer neuron exists. If, for example, an agent represents a wind farm with five turbines, the corresponding RNN will contain 20 neurons in its input layer: Wind speed, wind direction, nacelle orientation, and momentary power output for each turbine are fed to the RNN that will, in return, produce a single floating-point value that represents the farm's active power output (or consumption) at a time interval in the future. This time interval, $\tilde{t}_f$, is, by design, two interval lengths in the future from the journal entry that is the basis for the forecast. I.e., given $\tilde{t} = [t_1; t_2)$ as the time interval of the current journal log entry, the time interval of the forecast is:

$$\tilde{t}_f = [t_1 + 2(t_2 - t_1); t_2 + 2(t_2 - t_1)) \,. \qquad (5.3)$$

This two-intervals-length offset is chosen intentionally: It allows for the typical planning ahead and includes additional time in which the agents' communication and calculation stages may take place. It has a minimum length of 10 min, which is arbitrarily chosen, but based on the 10-minute interval used to calculate means in meteorological data.

In addition to the different value domains, the RNN and the `JournalEntry` use two different concepts of time: Any journal entry carries the timespan with it; it has therefore an explicit notion of time. The RNN, through its context layer, offers an implicit notion of time. More specifically, no information on time and date are fed to the RNN. This is intentional: Any information about date, time, or season, do not enhance the pattern, but rather degrade its value, because the pattern loses its generality. Consider, for example, the following input-to-output mapping:[8]

$$(0.9699, 0.4589, 0.3316) \mapsto (0.9697) \ .$$

The RNN will be trained to recognize this pattern and variances thereof. If we add, for example, the day of the year to the pattern, we obtain:

$$(0.9699, 0.4589, 0.3316, 0.2158) \mapsto (0.9697) \ ,$$

which would be valid for a reading on a spring day.[9] The pattern introduced above can naturally repeat itself[10] on a day in fall. However, if presented to the ANN, the following vectors effectively form a different pattern:[11]

$$(0.9699, 0.4589, 0.3316, 0.8225) \mapsto (0.9697) \ .$$

In order to generalize, the RNN can be trained to recognize the first pattern, given different values for the day-of-the-year position. However, when inspecting the resulting weights in the ANN, it becomes obvious that the network has effectively learned to *ignore* the input corresponding to the date. Thus, any explicit time information is excluded; the notion of time is sufficiently represented by the RNN's context layer that remembers series of input vectors. For this reason, the forecaster maintains a *sliding window* of journal entries for feeding to the RNN.

Each forecast consists of:

1. A series of journal entries, converted to floating-point vectors, that represents a short-term history leading to the current state

2. the entries that form the basis for the current forecast.

The short-term history is fed to the RNN in order to initialize its context layer to the current state. This attributes to the network's different concept of

---

[8]The actual contents of the input and result vector are insignificant for the argument, but their meaning is given here for the sake of completeness. The input vector represents the state of a wind turbine as a triplet of the nacelle position, the current wind speed, and the momentary power output; the output vector contains the forecasted power output. NB. that the input vector is also intentionally kept incomplete.

[9]$366 \cdot 0.2158 \approx 79$; the 79th day of the year is March 19th in a regular, i.e., non-leap, year.

[10]Not exactly, obviously, but the second pattern can occur with a negligible variance that we can treat the two as being the same.

[11]$0.8225 \cdot 366 \approx 300$ corresponds to October 26th.

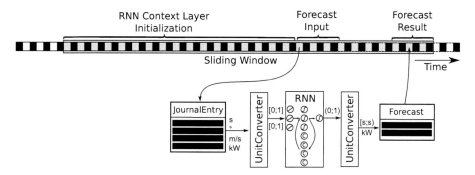

Figure 5.2: The sliding window structure maintained in the forecaster module

time: Whereas a journal entry carries the date/time information explicitly, the RNN must first 'initialize its memory.' Fig. 5.2 illustrates the sliding window and the data pipeline each piece of information travels through during the creation of a forecast.

A number of `JournalEntry` objects must therefore form the input of the ANN. The `UnitConverter` decouples the agent's classes from the ANN. Thus, it addresses the issue of technical debt that can arise when an ANN is an integral part of a piece of software that is tightly coupled with the ANN, as mentioned in Section 2.4 and especially by Sculley et al. (2014). Finally, these unit converters are used to convert the output of the ANN back to another type of object, the `Forecast`. This is used to calculate the node's power equilibrium or disequilibrium and serves as data structure during the agents' communication.

Training of the Forecaster's Artificial Neural Network

Not only at the beginning, but also during the agent's lifetime must the forecaster maintain its accuracy. The RNN is, therefore, continuously subject to training in order to retain its accuracy with regards to changing input patterns. However, simply using all available journal entries for training is not only ineffective, but also dilutes the RNN's ability to generalize due to overfitting of a particular class of patterns.

This problem is easily solved by comparing vectors and selecting only those that are significantly different from each other. Oftentimes, this training set is hand-selected in order to ensure that the training set has sufficient coverage.[12]

---

[12]Cf., e.g., Maqsood et al. (2004).

For the Universal Agent proposed in this thesis, this is not an option: It would contradict its deploy-and-forget maxim. The agent must therefore select the relevant patterns on its own.

The forecaster module must, therefore, calculate the similarity of two patterns as well as requiring a notion of when a pattern is relevant for training or not. This is obviously a two-stage process.

First, in order to compute the similarity of two vectors, their norm is calculated. Specifically, the Universal Agent uses the ℓ^2-norm:

$$\mathrm{d}(\boldsymbol{p}, \boldsymbol{q}) = \sqrt{\sum_{k=1}^{|\boldsymbol{p}|} (p_k - q_k)^2}, \quad |\boldsymbol{p}| = |\boldsymbol{q}| \ . \tag{5.4}$$

This allows us to define the equality of two patterns in terms of representing a *class* of patterns:

Definition 5.1. *Two vectors representing patterns, $\boldsymbol{p}$ and $\boldsymbol{q}$, are considered to belong to the same class of patterns and therefore equal, with regards to the training of the forecaster's RNN, if their Euclidean distance is smaller than the defined cutoff value, ε:*

$$\boldsymbol{p} = \boldsymbol{q} \quad \Leftrightarrow \quad \mathrm{d}(\boldsymbol{p}, \boldsymbol{q}) < \varepsilon \ . \tag{5.5}$$

In Fig. 5.3, the cutoff value is $\varepsilon = 0.4$, shrinking the training set considerably.

Using this equality operator, we can introduce a `PatternSet` class representing a set of pattern classes. With regards to the set's size, we have two options:

1. On nodes with sufficiently sized hardware with regards to storage and computing performance, the set's size does not necessarily need to be bounded.

2. Smaller nodes using, for example, embedded systems, will need to place a limit on the pattern set's size in order to economize their resources.

Usually, the second option represents the more sensible choice, especially since no system features unlimited resources, and devices running unattended will exhaust their memory at some point if no limit is enforced.

Additionally, the size of this `BoundedPatternSet` is given naturally through the size of the RNN that is used for forecasting. Recall that Section 2.4 offered an estimation of the number of distinct patterns required for training in Eq. (2.10):

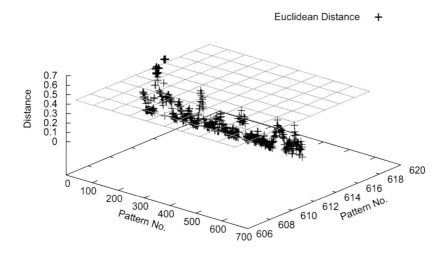

Figure 5.3: Euclidean distances of patterns derived from the reference situation

$$n = |\boldsymbol{w}| \log |\boldsymbol{w}| \ . \tag{5.6}$$

When the `BoundedPatternSet` is trimmed to its optimal size—i.e., super-fluous patterns are removed—, the class selects those patterns for removal that have a low number of occurrences. Therefore, the bounded pattern set is a sorted set, where its members—the patterns—are ordered by their number of occurrences. Every time a pattern is added to the set, the `BoundedPatternSet` checks whether it is already contained in the set or not, using Eq. (5.5) in Definition 5.1. If it is, the pattern's occurrence value is increased instead of storing the pattern again.

When the agent instance is initially set up, neither a set of patterns nor a trained RNN is available. In order to follow the deploy-and-forget philosophy, the agent must remain in a disconnected or simulation mode in which it does not control the node, but tunes the forecaster module.[13] The RNN's

---

[13]Of course, a vendor can offer to provide pre-trained and pre-tuned RNNs or, in general, ANN configurations. This would, for a third party, constitute a way to generate additional revenue.

size is then continuously adjusted to meet the error threshold supplied by the agent's constraints module.[14] When calculating the ANN's prediction error during training using *Mean-Squared Error* (MSE),[15] a target MSE value can be meaningfully derived from the allowable error the Agent's constraints calculator[16] provides: The RNN has one output neuron that denotes the forecasted power value, the number of training patterns is also known from the size of the pattern set, $|P|$. Since the MSE of the individual training sets is added to arrive at the total MSE and the maximum power generation or consumption of a node is known from the corresponding `UnitConverter` instance, deriving the maximum allowable MSE is a simple calculation.

The training of RNNs is, through their recurrent nature, more complex than the training of a simple, feed-forward ANN. Since the result of a previous run influences the result of the current calculation done by the RNN, the error landscape of an RNN training pass provides more local minima. A possible technique that forms a solution is therefore the stochastic approach, for example, through a genetic or evolutionary algorithm.[17]

This thesis employs a training algorithm that combines its evolutionary nature with a deterministic, gradient-based approach. This algorithm, the multipart evolutionary training algorithm codenamed 'REvol' and conceived by Martin Ruppert, has until now been published only once (Ruppert et al., 2014). Section 5.2, which follows, will therefore analyze the algorithm's internals and present its inner workings more extensively as the original publication.

5.2   The Multipart Evolutionary Training Algorithm for Artificial Neural Networks

Object and Population

The training algorithm employs an approach similar to the classic evolutionary algorithms in that it defines a number of individuals that live in a population.

---

[14]Cf. Section 4.1.

[15]If a vector $\boldsymbol{q}$ is predicted with a vector $\hat{\boldsymbol{q}}$, the MSE is:

$$\mathrm{mse}(\boldsymbol{q}, \hat{\boldsymbol{q}}) = \frac{1}{|\boldsymbol{q}|} \sum_{k=1}^{|\boldsymbol{q}|} (\hat{q}_k - q_k)^2 \ .$$

[16]Cf. Fig. 4.2.

[17]The background and other possible approaches have been outlined in Section 2.4.

In contrast to the typical approach, an object[18] consists of two vectors: A parameter vector and a scatter vector, each with the same size.

The object's parameter vector corresponds to an individual's genetic string and represents a possible solution to the problem that is being solved, for example, the weights of an ANN. Each component of the scatter vector limits the variability of the corresponding parameter vector's component. Finally, an object also encapsulates the individual's remaining *Time To Live* (TTL):

$$o = (\boldsymbol{o_p}, \boldsymbol{o_s}, ttl) , \qquad\qquad\qquad ttl \in \mathbb{Z} , \qquad (5.7)$$

$$\boldsymbol{o_p} = (o_{p_1}, o_{p_2}, \ldots, o_{p_k}, \ldots, o_{p_n}) , \qquad \forall o_{p_k} \in \boldsymbol{o_p} : o_{p_k} \in \mathbb{R} , \qquad (5.8)$$

$$\boldsymbol{o_s} = (o_{s_1}, o_{s_2}, \ldots, o_{s_k}, \ldots, o_{s_n}) , \qquad \forall o_{s_k} \in \boldsymbol{o_s} : o_{s_k} \in \mathbb{R} . \qquad (5.9)$$

Several parts of the algorithm require random numbers, as would be expected from an evolutionary algorithm. However, different distributions are used, based on drawing a random number from a bounded, uniform distribution, $X \sim \mathcal{U}[0;1)$. In the following paragraphs, this will be shortened to $X_k^{\mathcal{U}[0;1)}$ with $k = 1, 2, 3, \ldots, n$, where each index identifies a separate drawing.

The generation of the initial population depends on a user-supplied base object, the *origin object*, o_0. More specifically, all objects living in the initially population are derived from this base object. First is a derived object's scatter vector is generated from the origin object's scatter vector. Then, the algorithm calculates the derived object's parameters vector using the origin object's parameters vector and the previously generated, derived scatter vector:

$$o'_{s_k} = o_{0,s_k} \cdot \exp\left(0.4 \cdot \left(0.5 - X_1^{\mathcal{U}[0;1)}\right)\right) , \qquad (5.10)$$

$$o'_{p_k} = o_{0,p_k} + o_{i,s_k} \cdot \left(X_2^{\mathcal{U}[0;1)} - X_3^{\mathcal{U}[0;1)}\right) . \qquad (5.11)$$

From Eq. (5.10), we know that any derived object's scatter vector will initially contain values in the interval $\left(o_{0,s_j} \cdot e^{-0.2}; o_{0,s_j} \cdot e^{0.2}\right] \approx \left(o_{0,s_j} \cdot 0.82; o_{0,s_j} \cdot 1.22\right]$. Eq. (5.11) defines an object's parameters initially to be based on the base object's parameters and lie within its own scatter. The drawings of a random number from the uniform distribution create a distribution that resembles a triangle distribution with the interval $(-1; 1)$ and the center $c = 0$. Hence, the generation of the initial population favors a placement near the origin object, with the population thinning out the greater the distance from the origin (o_0).

---

[18]Throughout the next sections, the terms *individual* and *object* are used interchangeably.

Consider, as an example, Ackley's function (Ackley, 1987; Bäck, 1996), plotted in Fig. 5.4:

$$\text{ackley}(\boldsymbol{x}) = -a \cdot \exp\left(-b \cdot \sqrt{\frac{1}{d} \sum_{k=1}^{d} x_k^2}\right) - \exp\left(\frac{1}{d} \sum_{k=1}^{d} \cos(cx_k)\right) + a + \exp(1.0) \, .$$

$$(5.12)$$

Ackley's function is commonly chosen because it offers a hypercube with many local minima and maxima an optimization algorithm needs to escape. Especially the classic *Hill Climbing* algorithm is easily trapped in one of the local minima. It has its optimum at $\text{ackley}(\boldsymbol{x} = (0, \ldots, 0)) = 0$. The parameter d specifies the number of dimensions; in order to allow plots of this function be made, we set $d = 2$. Recommended values for the other parameters are $a = 2$, $b = 0.2$, and $c = 2\pi$.[19]

If we choose the origin object's vectors as $o_{0,p} = (5.0, 5.0)$ and $o_{0,s} = (10.0, 10.0)$, we can observe in Fig. 5.4 how the algorithm places the initial population around the base object. If we lower the values of the scatter vector's components, the population will be generated closer to the origin object; larger scatter values increase the spread.

Algorithm Main Body

After the generation of the initial population, the algorithm follows the general design of other evolutionary algorithms: Until one object satisfies the target condition or the number of iterations reaches a predfined maximum, the algorithm generates a new individual, uses a fitness function to evaluate it, and possibly enhances the population with it.

However, there are several points where it diverges from the plain approach. The algorithm has an explicit notion of success that is not simply bound to an object being the new best one: The population's overall success is represented as a real number. This success value is to be understood as the current mean success rate, representing the population's success in a given time frame. The algorithm tries not only to optimize the fitness function, but also how the population obtains a success.

A success is achieved when an object becomes the new best one. Thus, the algorithm can influence (with the goal to optimize) the success rate of

---

[19]These parameters are specific to Ackley's function, cf. Ackley (1987); Bäck (1996).

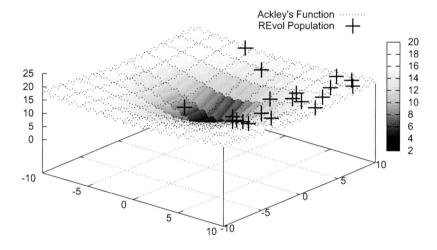

Figure 5.4: Plot of the initial population in Ackley's Function

the population of a number of iterations through the placement of individuals, i.e., by dynamically adjusting the shape of the *Probability Density Function* (PDF) during runtime. The parameter-wise static PDF we have encountered in Section 5.2 becomes a *dynamic reproduction probability density function*. We will encounter its actual specification in Section 5.2, which follows; here, in the main loop that is shown in Algorithm 2, it is the influencing variable set.

In order to avoid fluctuations in this variable—*success*—, it is dynamically averaged over a fixed number of iterations, represented by the user-supplied variable T. The averaging is done by a time-discrete PT1 element:

$$\text{pt1}(y, u, t) = \begin{cases} u & \text{if } t = 0, \\ y + \dfrac{u - y}{t} & \text{otherwise,} \end{cases} \tag{5.13}$$

with $t = T$ being always true in the context of the multipart evolutionary algorithm.

Algorithm 2 Main loop of the multipart evolutionary algorithm

---

procedure REVOL(o_0, $targetError$, k_{max}, FITNESS)
  global $populationSize$, $eliteSize$, $targetSuccessRate$, $maxNoSuccess$, T
  local i, $error$, $lastSuccess$, O, $success$
  $k \leftarrow 0$
  $lastSuccess \leftarrow 0$
  $O \leftarrow$ GENERATEINITIALPOPULATION(o_0, $populationSize$)
  for all $o \in O$ **do**
    $o_{fitness} \leftarrow$ FITNESS(o)
  end for
  $success \leftarrow targetSuccess$
  while $error > targetError \wedge k < k_{max} \wedge k - lastSuccess \leq maxNoSuccess$
    do
    $o' \leftarrow$ GENERATEOBJECT(O, $success$)
    $worst \leftarrow O_{|O|}$
    if $o' < worst$ **then** ▷ New object is better than the current worst one.
      if $worst_{Age} \geq 0$ **then**
        $success \leftarrow$ pt1($sucess$, 1.0, T)
      else
        $success \leftarrow$ pt1($sucess$, -1.0, T)
      end if
      $O \leftarrow O \setminus \{worst\}$
      $O \leftarrow O \cup \{o'\}$
    end if
    $O \leftarrow$ SORT(O)   ▷ Sort so that the best object is the first one.
    if $O_1 = o'$ **then**
      $o'_{age} \leftarrow k$
    end if
    for all $o \in \boldsymbol{O}$ **do**
      $o_{age} \leftarrow o_{age} - 1$
    end for
    $success \leftarrow$ pt1($success$, 0.0, T)
    $k \leftarrow k + 1$
    $error \leftarrow O_{1,fitness}$
  end while
  return O_1
end procedure

Generating Individuals

For the generation of a new object, the algorithm chooses an object from the elite o^+ and another one from the general population o^-. The elite is contained in the general population: *Elite $\subset$ Population*. Therefore, the object denoted by o^- might as well be a member of the elite. The algorithm then determines the value of the two influencing factors that apply during the creation of a new object and that form the distinct features of this algorithm:

1. The current rate of success

2. the implicit gradient information.

Remember that the current rate of success influences the spread, i.e., the area within which a new object can potentially be placed. However, this area does not feature a uniform PDF. Instead, the placement of the two objects chosen relative to each other is used to calculate the implicit gradient, i.e., the direction within which a newly generated individual must be placed with a high probability in order to reach the minimum. These two factors carefully balance each other: Putting a strong emphasis on the implicit gradient information would turn the multi-part evolutionary strategy into a 'poor man's gradient decent,' whereas a high influence of the dynamic reproduction probability density spread will make the algorithm lose its orientation.

The multipart evolutionary training algorithm defines the current rate as success as follows:

$$successRate = \frac{success}{targetSuccess} - 1.0 \ . \tag{5.14}$$

The success rate already influences the use of the implicit gradient thus:

$$xlp = \left(\sum_{k=4}^{14} X_k^{\mathcal{U}[0,1)} - \sum_{k=15}^{20} X_k^{\mathcal{U}[0;1)} \right) \cdot w_G \cdot \exp\left(w_G \cdot successRate \right) \ . \tag{5.15}$$

The concatenation of random number drawings can be approximated by a drawing from a normal distribution, $X \sim \mathcal{N}\left(-2.0, \frac{16}{12}\right)$.[20]

An additional factor of 0.5 is applied when $xlp > 0$, which favors a decent more strongly than an ascent in addition to the mean of the approximating

---

[20]The mean of the PDF is located at -2.0, and the variance of each drawing $X \sim \mathcal{U}[0;1)$ is $\frac{1}{12}$. We can describe the actual PDF as the convolution of each drawing's PDF, i.e., if

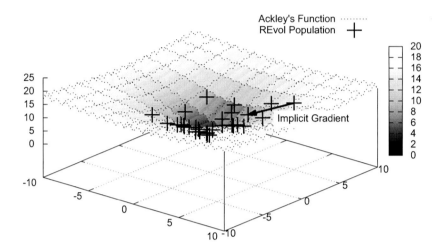

Figure 5.5: Implicit gradient in a REvol population

normal distribution ($\mu = -2$). The parameter w_G is the gradient weight. Higher values allow a faster decent, setting $w_G = 0$ disables the influence of the implicit gradient entirely. Sensible values range from 1.0 to 3.0; $w_G > 3.0$ will diminish the influence of the dynamic reproduction probability density function as to make it useless and is therefore not recommended. A piece of implicit gradient information between two objects is delineated in Fig. 5.5.

The multipart evolutionary strategy continues to create the new object's scatter and parameters vector. It does so by first updating the elite object's scatter vector, since this naturally influences the new object's scatter:

$$o_s^+ = o_s^+ \cdot \exp(w_S \cdot successRate) \ . \tag{5.16}$$

$X = X_1 + X_2,$

$$f_X(x) = \int_{-\infty}^{\infty} f_{X_1}(x - x_1) f_{X_2}(x) \ dx \ .$$

Several concatenations can be constructed as $X' = X_1 + X_2$, then $X = X' + X_3$. Calculating the coefficients of the polynomials that are the result of the necessary convolutions is out of the scope of this thesis.

Note that Eq. (5.16) uses the success rate calculated in Eq. (5.14): The dynamic reproduction probability density is introduced to the elite object's scatter, modifying the vector of the implicit gradient information later. With the elite object's scatter vector updated according to the current success, the new individual's scatter is generated:[21]

$$o'_{s_k} = \frac{1}{2} \left(o_{s_k}^+ + o_{s_k}^- \right) \cdot \exp \left(X_{21}^{\mathcal{U}[0;1)} - X_{22}^{\mathcal{U}[0;1)} \right) . \qquad (5.17)$$

With a new scatter value available, the algorithm can finally generate a new parameter set. The corresponding scatter value serves to limit the new parameter relative to the other object's parameter values:

$$o'_{p_k} = o'_{s_k} \left(\sum_{n=23}^{27} X_n^{\mathcal{U}[0;1)} - \sum_{n=28}^{32} X_n^{\mathcal{U}[0;1)} \right) + xlp \left(o_{p_k}^+ - o_{p_k}^- \right) + o_{p_k}^+ . \qquad (5.18)$$

Here, the repeated drawings $X \sim \mathcal{U}[0;1)$ create a PDF similar to $\mathcal{N}\left(-2.5, \frac{10}{12}\right)$.

Observe how the implicit gradient information is used in Eq. (5.18) in the term $o_{p_k}^+ - o_{p_k}^-$. The last term of the sum constitutes the direction of the gradient vector, i.e., in direction of the elite object. Note how, via the variable xlp from Eq. (5.15), the dynamic spread is involved in generating the parameters of the new individual.

From Eqs. (5.14), (5.15), and (5.18), we can follow the influence of the dynamic reproduction probability density that can be understood as a function of the current population's state: If the current representation of success is greater than the target success, i.e., *success > targetSuccess*, the population's spread increases, with the intensity represented by the user-supplied success weight, w_S. For *success < targetSuccess*, the population is drawn together.

This feature is important when escaping local minima purposefully. Remember the main portion of the algorithm outlined in Algorithm 2: The pt1 function dynamically averages the current success rate. If we tune the fitness function in our example that the algorithm continues to run even when the global minimum is reached, we can observe that the population, which previously had converged in the global minimum, begins to escape from the minimum in search of a better minimum. This is shown in Figs. 5.6a and 5.6b.

Eqs. (5.14) and (5.15) and Algorithm 2 reveal that the population size, the individuals' initial TTL, and the success rate are connected. The pt1(y, u, t)

---

[21]The index k indicates separate drawings of random numbers for each member of the respective vector.

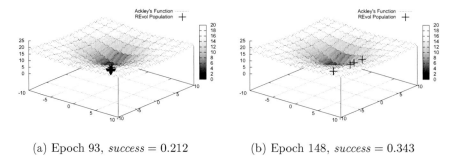

(a) Epoch 93, $success = 0.212$ (b) Epoch 148, $success = 0.343$

Figure 5.6: REvol tuned to let the population escape a minimum

function is only then called, with $u = 1.0$, when a new object is generated that satisfies the following features:

1. It is better than the current worst one

2. it replaces a still living individual (i.e., $ttl > 0$).

If the population is big and the initial TTL values small in comparison, the new objects will only replace dead individuals after a certain number of iterations, letting the success rate drop ($success \ll targetSuccess$). The population will then get indefinitely stuck in the current minimum. This is the algorithm's Achilles heel. It is therefore advisable to set the start TTL in correspondence with the population size: $ttl \geq 5|O|$.

Ruppert's algorithm does not rely solely on the implicit gradient information; it still is an evolutionary strategy in which standard crossover happens. Ruppert et al. (2014) present this specific portion of the algorithm in pseudocode.

5.3   Forecasting Accuracy and Efficiency

Experiment Configuration

We will now, following the analysis in the previous sections, set up the multipart evolutionary algorithm to prove its worth in a real-life situation against a well-understood competitor. This experiment utilizes data from the reference situation that was outlined in Section 3.2 and forecasts the output of the 'Bare Hill Wind Farm'.

The experiment will utilize an Elman RNN, as proposed in Section 5.1. The original introduction of the multipart evolutionary algorithm compared the algorithm to SA,[22] which showed a better performance of the presented algorithm in comparison with standard SA. While SA is, basically, a stochastic informed search algorithm in the same way as the multipart evolutionary algorithm is, too, it does not use the concept of a population whose individuals represent candidate solutions. Instead, SA only keeps one candidate solution it modifies over a number of cycles.[23]

The evolutionary nature of the presented algorithm is better matched by other algorithms that employ similar mechanics. A central concept of the multipart evolutionary algorithm is the influence individuals have on each other, not only in terms of the classical crossover, but also the distance relative to its parents an offspring is created in.[24] The population's overall success as well as the implicit gradient information the position of the parents delivers is incorporated in the offspring's parameters as well.[25]

PSO employs mechanics that stem from a similar idea. Here, the population is a *swarm* of *particles*. Each particle has a certain position, which is a vector of parameters to the fitness function. Additionally, a particle also features a velocity. On any iteration, a particle's new position is computed by combining its velocity with its current position and taking the particle's neighborhood in account. Of PSO, a number of variations exist. In this comparison, we use *Standard Particle Swarm Optimization* (SPSO) in its 2011 revision. The details of the swarm's initialization, the definition of a particle's neighborhood, and the combination method to compute a particle's new position, are described by Clerc (2012).

In order to aid the comparison, parameters are chosen in similar ranges wherever possible. E.g., the swarm size of SPSO is usually set to 40 particles, thus the population size of the multipart evolutionary algorithm is also set to 40 individuals. Each algorithm is allowed to run for the same number of iterations. Table 5.1 summarizes the relevant parameters of the two training algorithms as they were used in this comparison.

The overall result that is expected from this experiment is to verify an agent's ability to forecast the power balance at the local node it represents. An hour-ahead forecast is sufficient to allow all agents to enter the distributed planning phase and form short-term contracts to counter a prognosticated

---

[22]Cf. Ruppert et al. (2014).
[23]Cf. Kirkpatrick et al. (1983).
[24]Cf. Eqs. (5.17) and (5.18).
[25]Cf. Eqs. (5.14) and (5.15).

Table 5.1: Parameters of REvol and Standard Particle Swarm Optimization 2011

Parameter	REvol Setting	Parameter	SPSO Setting
Population Size	40	Swarm Size	40
Elite Size	4		
Max. Epochs	50 000		
Max. Epochs without success	10 000	Max. Epochs	50 000
T	5000		
Gradient Weight	0.8	C	$\frac{1}{2} + \ln 2$
Success Weight	1.1	W	$\frac{1}{2\ln 2}$

disequilibirum. Revisiting Fig. 5.2, we can see that the agents have at least 10 min for this.

While SPSO is able to use a concept similar to the evolutionary algorithm's implicit gradient information through the PSO's concept of a *particle neighborhood*, it cannot purposefully escape a local minimum through a measurement equivalent to the evolutionary algorithm's success rate. It relies on the particle neighborhood, their recorded best and previous best positions to draw a particular particle to a minimum. SPSO relies on these neighborhoods to find the best minimum, i.e., each particle neighborhood occupies a minimum and the best neighborhood wins the search. Therefore, the expected outcome is that the multipart evolutionary algorithm escapes these minima and yields a better overall training result than its contender.

Results

Supervised training of an ANN is made up of two stages, which also create a partition of data: First, the training of the network, second, the verification of the training's success with the second part of the data set. The forecaster's sliding window concept of the data pipeline along with the pattern store creates this partition naturally.

When comparing the training performance, one can suspect from the algorithm's designs alone that the multipart evolutionary training algorithm will arrive faster at a result than the SPSO. Indeed, the SPSO required several orders of magnitude more time than the evolutionary algorithm. However, even

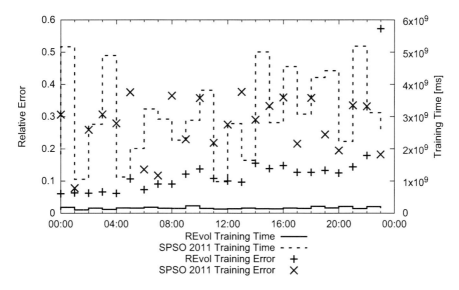

Figure 5.7: Training algorithm performance

though the SPSO modifies and re-evaluates all particles for a given epoch—as outlined in the previous section—, it was not able to achieve a better training result. With few exceptions, the final weight configuration created by the SPSO was inferior to that created by the multipart evolutionary algorithm, as far as the final training error was concerned. Fig. 5.7 displays a typical number of results for both algorithms over the course of 24 hours.[26] Although the timing results, being absolute numbers, are dependent on the hardware and build configuration,[27] the graph serves to illustrate the relative difference in training duration.

The activation function chosen and its configuration particularly influences the duration of the training. Recall Section 2.4 that cited the discussion of Jordan (1995) on why the logistic function is often the most beneficial. Fig. 5.8 compares the power curve of a wind turbine with the $\tanh(x)$ function and

---

[26]Fig. 5.7 displays the training of the same timespan that is depicted in Fig. 3.3, plotted as one-hour means to improve readability.

[27]The tests were run on a machine powered with two Intel® Xeon® 5140 *Central Processing Units* (CPU) running at 2.33 GHz; the C++ test code compiled with GCC version 4.8.5 using the compile flags -O3 -march=native.

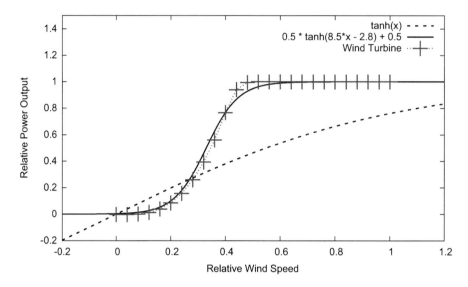

Figure 5.8: The Artificial Neural Network's activation function and the power curve of a wind turbine

offers a modified activation function alongside:

$$\mathrm{af}(x) = 0.5 + \frac{1}{2}\tanh(8.5x - 2.8) \ . \tag{5.19}$$

The domain-specific knowledge that lies in the shape of the wind turbine's power curve aids in the training of the ANN and reduces training time by 17.2 %.

Weather is a complex phenomenon; thus, there is no 'one network size to rule them all.' If the RNN contains too many connections, it will overfit, i.e., lose its ability to generalize and derive meaningful results from yet unknown, but similar to already learned input patterns.[28] However, the RNN's size is also its memory, and the number of connections might as well be too small, therefore unable to model the situation underlying the forecasts. The size of the RNN—i.e., the number of neurons and the number of connections, $|\boldsymbol{w}|$—must be dynamically adjusted during runtime. A mechanism to dynamically grow and shrink the RNN can be as follows:

---

[28] Cf. Section 2.4.

1. Initially, size the RNN's to contain 1.5 times as many neurons as the input layer contains. Fully connect each layer.[29]

2. Train the RNN:

   a) If the resulting final error of the training is equal to or below the required training error as indicated by the agent's constraints module,[30] the Forecaster applies the Optimal Brain Damage algorithm[31] to ensure that the RNN does not overfit.

   b) If the resulting training error is not satisfying, the RNN is first checked to be fully connected. If it already is, the size of the hidden layer is doubled. A new training process is then started.

This algorithm is depicted in Fig. 5.9.

While the dynamic adjustment of the RNN's size helps to capture more complex weather patterns, it can eventually also cause the network to grow to a size that is no longer suitable for calculation with the computing power present at the local node. However, any ANN can profit from specialized hardware that is able to compute a large number of floating-point operations in parallel, such as a *Field-Programmable Gate Array* (FPGA) can. FPGAs are not as expensive as high-performance CPUs or *General-Purpose Graphics Processing Units* (GPGPU) and also consume less power when operating. FPGAs have already been used to—and are, in fact, well known to—calculate ANNs and undertake their training (Lysaght et al., 1994; Zhu and Sutton, 2003; Omondi and Rajapakse, 2006).

The overall performance of the actual forecast, quantified through its *Mean Absolute Error* (MAE)[32] and *Root Mean Squared Error* (RMSE),[33] are outlined in Table 5.2. For better reference, these values are compared with the results achieved by Liu et al. (2012).

---

[29]Except where otherwise dictated by the underlying pattern, e.g., the Elman RNN pattern
[30]Cf. Section 4.1.
[31]Cf. Section 2.4.
[32]

$$\mathrm{mae}(\boldsymbol{q}, \hat{\boldsymbol{q}}) = \frac{1}{|\boldsymbol{q}|} \sum_{k=1}^{|\boldsymbol{q}|} |\hat{\boldsymbol{q}}_k - \boldsymbol{q}_k|$$

[33]

$$\mathrm{rmse}(\boldsymbol{q}, \hat{\boldsymbol{q}}) = \sqrt{\frac{1}{|\boldsymbol{q}|} \sum_{k=1}^{|\boldsymbol{q}|} (\hat{\boldsymbol{q}}_k - \boldsymbol{q}_k)^2}$$

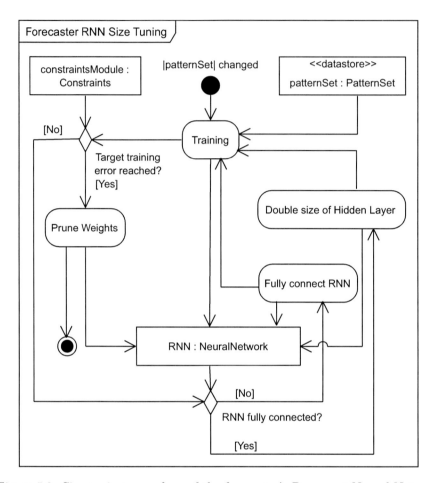

Figure 5.9: Size tuning procedure of the forecaster's Recurrent Neural Network

An excerpt of a forecasted day at the 'Bare Hill Wind Farm' node is depicted in Fig. 5.10.

While successful, pure node-local forecasting of power generation or consumption based on historic data alone will not, and, in fact, cannot be the ultimate solution to this problem. As long as a pattern class has not been

Table 5.2: Comparison of the Universal Smart Grid Agent's forecaster module's performance with known values from literature

Metric	Liu et al. (2012)	Agent Forecaster
MAE	14.149	7.13
RMSE	14.994	8.47

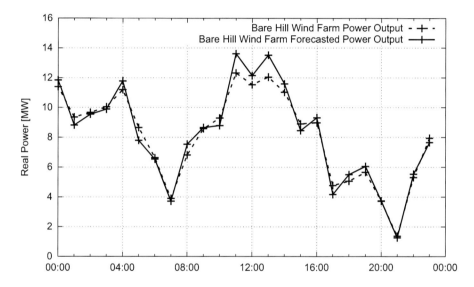

Figure 5.10: Result of node-local forecasting at 'Bare Hill Wind Farm'

encountered, the forecast must fail. Complex weather conditions might also lead to contradicting patterns. To conclude, node-local forecasting is a valuable addition to a global forecast and helps to account for the idiosyncrasies of an agent's environment. It cannot, however, replace forecasts created by national weather services.

6 Social Component: Inter-Agent Communication

6.1 Motivation

The fundamentals in Section 2.4 showed us that, for every agent, the ability to communicate and coordinate with its partner agents is an elementary, intrinsic property of it. The word 'protocol' itself, coming from the ancient Greek word 'protókollon,'[1] initially described something akin to today's envelope folder and contained bibliographic data. French diplomacy added a second meaning to the a posteriori concept of the original 'protókollon': A set of rules that should be followed in order to follow the protocol.

Thus, in the computer age, too, a protocol defines a set of behavioral rules as well as a set of data and their encoding. In the context of a multi-agent system, these behavioral rules gain special weight, since only a purposeful cooperation of all necessary agents, according to a set of rules yielding constructive behavior, will contribute to the formation of a solution. Therefore, a protocol for a multi-agent system should first define a set of rules and then deduce the data transmitted from what information is necessary for the receiver of a message to follow the intended ruleset. This information must be carefully selected, because it forms an extract of an agent's current model of its world that it communicates, i.e., it must communicate enough of its own state in order to help other agents to arrive at the necessary conclusions.

The nature of the multi-agent system is largely characterized by proactive behavior, as we know from Chapter 4. Recall Section 2.3 that presented protocols in use in the smart grid domain. Many define a query-based interaction, i.e., a host polls another host's state. While this is appropriate for sensor

---

[1]Ancient Greek προτόκολλον: 'prótos' means 'first,' and 'kólla' translates to 'glue.'

devices that are not sized resource-wise to actively participate in a networked communication, this active-passive relationship is not approporiate for the active-active relationship between agents.

The smart grid does not offer a homogeneous communications technology landscape, nor can it: Between remote wind farms, solar power plants, and centrally connected consumer groups in cities, no single technology is available to connect them all. Any protocol aimed at enabling the information interchange in a smart grid must therefore either settle for the common denominator,[2] or cope with a range of communication technologies.

The ISO/OSI stack model presented in Section 2.3 has been created in order to allow the exchange of protocols in one layer with the smallest possible impact on other layers. It is therefore prudent to utilize existing protocols that allow access to different physical media and implement necessary algorithms for transmission error reduction, or transmission security.

This thesis proposes[3] a lightweight protocol for a smart grid that has been designed to provide a set of behavioral rules for the universal smart grid agent, define the data necessary to adhere to those rules, and re-use existing, proven technology where possible: The *Lightweight Power Exchange Protocol* (LPEP).

6.2   Design Principles

Network Layout

This thesis presents a decentralized concept through the proposed multi-agent system, in which each agent stands for a certain node in the power grid, e.g., a power plant or a neighborhood. Stringently, the communication network of the agents should resemble the power grid. This is achieved by forming an overlay network: If a node in the power grid is physically connected to another, a digital connection must exist between the two agents representing these nodes. The network layer of the ISO/OSI stack introduces the end-to-end concept in computer networks; protocols like the TCP introduce the connection concept on the transport layer within the ISO/OSI stack model.[4] The agent software utilizes the *Internet Protocol* (IP) and the TCP to connect

---

[2]Section 2.3 also showed that settling for the common denominator means the re-implementation of many proven technologies, which often causes more problems than it solves.

[3]Based on Veith et al. (2013, 2014)

[4]Cf. Fig. 2.8.

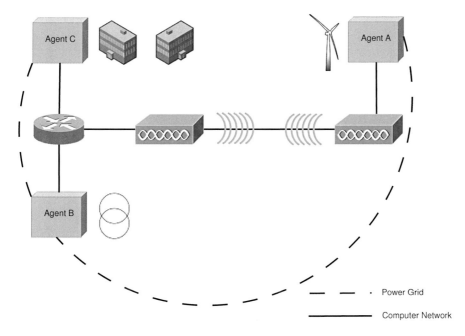

Figure 6.1: The power grid as an overlay network

two instances of the agent software, represented by network[5] addresses, with which the protocol then formulates a connection between two distinct agents. The result is a logic network layer residing on the application layer that overlays the existing communications infrastructure, thus modelling the power grid via a communications network. Creating overlay networks is a common technique in computer networking; the concept is depicted in Fig. 6.1.

The connections in this protocol concept are strictly end-to-end connections; multicasting must be realized on top of this concept by sending a message over several connections. Multicast classes and groups, such as the IP offers, are not implemented.

A power grid can be designed as a meshed structure in order to improve reliability. The resulting overlay network will therefore, too, form a mesh. Thus, several redundant paths can lead from one agent to another; hence, the LPEP requires all communication between agents to be *idempotent*. A request

---

[5]E.g., IPv6

for power generation that, through the meshed topology of the network, was duplicated, must not yield to two distinct responses, thus doubling the amount fed into the power grid: The results would be disastrous.

For the same reason, the LPEP incorporates a mechanism that limits the number of agents a message can pass through. A request that cannot be answered would otherwise potentially travel the network endlessly, becoming a 'zombie message' that would lead to congestion of the data network without benefit. Hence, the LPEP contains a TTL counter that is inspired by the IP.[6]

The undirected power grid does not offer any intrinsic information of its nodes and the overlay network follows suit. Consider the simple network of three connected agents $A_1 \leftrightarrow A_2 \leftrightarrow A_3$. A_1 and A_3 are independent of each other; neither of the two possesses any information about each other: Does A_1 constitute a power plant, a wind farm, or a factory? How is it connected? In fact, this information is not necessary and can even be interfering; nodes in the smart grid consume and generate power variably. A power plant can react to a surplus of power by throttling its turbines, therefore influencing the grid's power balance in a similar way as a consumer would do. However, the distributed Demand-Supply calculation proposed in this thesis must strive to minimize line loss and therefore needs an additional metric.

This metric is incorporated into the distance each message travels. This distance metric is based on the impedance of the power lines the message travels. A low distance value means that an agent nearby has sent the corresponding message. Thus, the structure of the power grid can be taken into account when calculating the solution set to a given situation of power disequilibrium. How the distance is calculated is part of the routing of messages and detailed in Section 6.2, starting on Page 128.

Message Types and Data Fields

The protocol exists in order to allow agents to communicate their local disequilibrium in terms of active or reactive power and to allow other agents to propose actions at their site in order to reach an equilibrium. Those messages must be sent before the actual situation of a disequilibrium becomes reality. Therefore, the communication takes place after a forecast has been made through the facilities described in Chapter 5.

In principle, agents send requests because they forecast a surplus of, or demand for, power at their position. Other agents act on this request by sending

---

[6]Cf. Section 2.3.

proposals for an increased consumption or increased generation at their site. A request for power is represented by a *demand notification*. Other agents can answer this demand by sending an *offer notification*.

Recall that the creation of message types and their data fields is motivated by a set of behavioral rules. This section will therefore introduce them in an abstract manner not tied to a specific implementation or encoding. We will use a tuple-like notation for messages, where individual fields are indicated by subscripts. For example, m_{id} denotes the ID of a message m.

The abstract[7] message data type defines the minimum fields that must be present in each transmission between two agents:

$$\text{Message} := (id, type, sender, receiver, is\ answer, answer\ to, \dots), \qquad (6.1)$$

where $id \in J$, $sender \in I$, $receiver \in I$, $is\ answer \in \{\text{true}, \text{false}\}$, and $answer\ to \in J$. These relations apply for all message types.

For any given exchange of these messages, it is necessary to unambiguously identify the agents that take part in this communication: Requests may be sent by any number of agents at a given time and thus the receiver of the message must be clear. The agent identifiers that are a part of the message must be unique at any given time. In addition, in order to formulate an answer to a specific message, the message itself must be unambiguously identifiable.

Since messages must be idempotent—a demand notification that was duplicated and arrives twice at an agent must not yield two distinct offer notifications—the message identifier is not only required to formulate answers, it also serves to ensure the idempotence of messages.

Facilities exist to derive identifiers that are unique[8] and can be created without a central system to issue them.[9] For agent addresses, a fixed scheme in the way of Ethernet MAC addresses (IEEE Standards Association, 2015) is possible. While this would still involve a central authority, the agent's address would be fixed once assigned and no service would need to be online at all times. Message identifiers, however, need to be generated on the fly during normal operations of an agent. A scheme like the *Universally-Unique Identifier* (UUID)

---

[7]Abstract in the same sense as *abstract class* is used in object-oriented programming: All data fields of the Message type are available in subtypes, but never is a pure instance of a Message type instantiated.

[8]At least, with a high probability

[9]This would introduce a single point of failure and defy the distributed, i.e., non-centralized, approach of the multi-agent system.

(Leach et al., 2005) therefore suggests itself. UUIDs could also be used for agent addresses, since no meaning needs to be contained in the agent identifiers.

In addition to the abstract data type defined in Eq. (6.1), both demand notification and offer notification, also need, obviously, to include the *amount of power* and the *type of power*[10] in units of Kilowatts, or KiloVArs, respectively. The amount of power transmitted is expressed as a closed interval, $\tilde{P} = [P_1; P_2]$, from which the receiver may freely choose a value. If $P_1 = P_2$, no choice is offered, obviously. If an agent can offer multiple choices, but only within distinct ranges, it must make several offers to the same request. The receiving agent must then collapse them into a single offer in order to preserve the exclusive-or semantic of the distinct messages. This allows the expression of both, infinitely adjustable, as well as partially variable or invariable, power generation or consumption, and accommodates, e.g. different load gradients in traditional power plants depending on the power output for to technical reasons.

Since a demand for or surplus of power is temporally bound, a *time interval*[11] must be included as well. Demand notification and offer notification therefore define a subtype of the abstract Message type defined in Eq. (6.1) thusly:[12]

$$\text{DemandNotification} <: \text{Message} \ , \tag{6.2}$$

$$\text{OfferNotification} <: \text{Message} \ . \tag{6.3}$$

They are furthermore defined as complete data types:

$$\begin{aligned} \text{DemandNotification} := \ & (id, \, type, \, sender, \, receiver, \, is\ answer, \, answer\ to, \\ & ttl, \, distance, \, timespan, \, answer\ until, \, value, \, power\ type) \ , \end{aligned} \tag{6.4}$$

where $type = 5$, $ttl \in \mathbb{N}$, $distance \in \mathbb{N}$, $timespan = [t_1; t_2)$, $value = [P_1; P_2]$, $P_1, P_2 \in \mathbb{N}$, and $power\ type \in \{\text{active}, \text{reactive}\}$. Except for the distinct value of *type* that denotes the message's type, these relationships apply to all other message types, too.

---

[10] Active or reactive

[11] If t_1 and t_2 denote points in time and are the bounds of the interval, a message always contains the interval $\tilde{t} = [t_1; t_2)$.

[12] The notation of a subtype, the *subtyping relation*, follows Abadi and Cardelli (1996).

$$\text{OfferNotification} := (id, type, sender, receiver, is\ answer, answer\ to,$$
$$ttl, distance, timespan, answer\ until, value, power\ type)\ ,$$
$$\text{(6.5)}$$

with $type = 6$. It becomes obvious that the existence of two distinct message types comes from a semantic reason, i.e., a question of behavior; their only difference data-wise is the value of the *type* field.

A demand notification expresses a shortage of power; an offer notification expresses a surplus of power. In the previous paragraph, we have discussed situations in which a forecast indicates a power shortage that is answered by an offer coming from a surplus of power. However, the reverse is also possible: That a forecast detects a surplus of power at some point in the future that can be consumed.[13] For example, a wind farm will generate more power when the wind speed increases if not curtailed, which is obviously due to the nature of the power generation. Thus, an offer notification can also initiate the communication.

In order to distinguish the two types of situation, the answer indicator[14] is used. An offer notification that is not an answer is, therefore, a request to the other agents to consume the surplus power. Stringently, the agent sending the offer notification will receive demand notifications that are answers. Likewise, a demand notification that is not an answer is a request to generate more power and is met by offer notifications that are answers.

When formulating its response, an agent is not required to exactly match the time interval or power value of the request: Smaller values are also possible. I.e., $Response_{value} \leq Request_{value}$ and $Response_{timespan} \subseteq Request_{timespan}$. The requesting agent's task in finding a solution is thus to select those responses that form the optimal (or good enough) solution to its situation of power disequilibrium.

After having received responses to its request, the agent begins to calculate a solution set that solves its current situation. Once a solution set exists, the agent must notify its partners that it takes them up on their offer (or demand) and therefore sends an *acceptance notification* to each agent whose response is in the solution set. The amount of power accepted from the offer is also

---

[13] A coal or gas power plant can also 'consume' power, i.e., influence the grid's power balance towards a power shortage by simply generating less power. Although it does not actually consume power, such an action may also constitute a demand notification as an answer to an offer notification.

[14] The message field that indicates whether a message is an answer or not

communicated herein; it must lie in the acceptable range of the offer. It does not need to deny responses: All unanswered responses are interpreted as denial.

$$\text{AcceptanceNotification} <: \text{Message} \, , \tag{6.6}$$

$$\begin{aligned} \text{AcceptanceNotification} := (\mathit{id}, \mathit{type}, \mathit{sender}, \mathit{receiver}, \\ \mathit{is\ answer}, \mathit{answer\ to}, \mathit{ttl}, \mathit{accepted\ value}) \, , \end{aligned} \tag{6.7}$$

where $\mathit{type} = 7$, $\mathit{is\ answer} = \text{true}$, and $\mathit{accepted\ value} \in \mathbb{N}$.

Once the responding agent receives the acceptance notification, it must answer with an *acceptance acknowledgement notification*. This concludes the four-way handshake and forms a short-term contract valid for the interval given in the response; Fig. 6.2 depicts a successful four-way handshake. The acknowledgement message is necessary because an agent may withdraw its demand or offer, for example, when a forecast is no longer valid and must be corrected. A *withdrawal notification* must be sent whenever data in a previous demand or offer notification becomes invalid.

$$\text{AcceptanceAcknowledgementNotification} <: \text{Message} \, , \tag{6.8}$$

$$\text{WithdrawalNotification} <: \text{Message} \, . \tag{6.9}$$

The two Message types carry the following data:

$$\begin{aligned} \text{AcceptanceAcknowledgementNotification} := (\mathit{id}, \mathit{type}, \mathit{sender}, \mathit{receiver}, \\ \mathit{is\ answer}, \mathit{answer\ to}, \mathit{ttl}) \, , \end{aligned} \tag{6.10}$$

where $\mathit{type} = 8$ and $\mathit{is\ answer} = \text{true}$, and

$$\text{WithdrawalNotification} := (\mathit{id}, \mathit{type}, \mathit{sender}, \mathit{receiver}, \mathit{is\ answer}, \mathit{answer\ to}, \mathit{ttl}) \, , \tag{6.11}$$

with $\mathit{type} = 9$ and $\mathit{is\ answer} = \text{true}$.

Power delivery (or consumption) is not instantaneous, especially with regards to traditional, steam-based power plants that must follow a certain load gradient.[15] However, the receiving agent does not have—and, in fact, does not need—any knowledge about the other nodes. The load gradient is therefore a part of the message itself in the form of the *answer until* field that carries a timestamp[16] that indicates at what time, at the latest, an answer may arrive at

---

[15] Cf. Sections 2.1 and 3.1 for details.
[16] Cf. Appendix B.2 for the exact type.

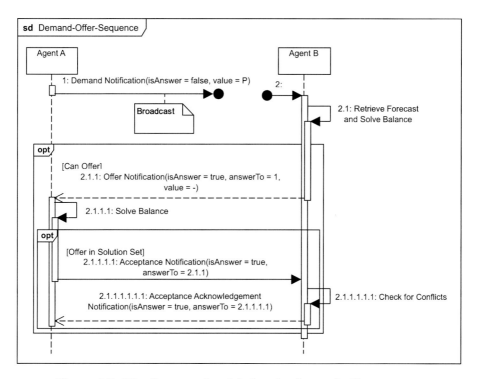

Figure 6.2: The four-way handshake of a demand-offer sequence

the sender. If the answer arrives afterwards, the responder must not act, i.e., the response becomes void.

Messages can arrive at any given time before the time indicated by the *answer until* field; the agent can therefore either wait until the last possible moment or calculate the solution set several times: The *answer until* values of the responses it receives may lie before the requester's own deadline.

Timing is obviously of importance here, since two factors influence the time-bound validity of a four-way handshake:

1. The computation time necessary for an agent to calculate a solution set

2. the network delay that forms the travel time of all messages.

An agent can measure the time it takes to finish solving the balance and use this as an estimate for future calculations. The four-way handshake yields several round trips: Four at a minimum, but most likely more, based on what communication protocols are used to connect the two agents. While a closed loop can be used to find exactly the right time to send acceptance notifications, this thesis does not propose one; in fact, measuring and predicting network delays is, as the TCP shows, a topic of research in itself (Tsang et al., 2003; Ha et al., 2008). Instead, a fixed time of 10 minutes to search for a solution set and answer with acceptance notifications is proposed here.[17]

The agent normally assumes that it can freely distribute power. However, line capacity, voltage drop, or other constraints may be a limit to this assumption. Therefore, agents must signal that a potential path in the power grid is not viable. They do so using a *constraint notification*:

$$\text{ConstraintNotification} <: \text{Message} \, , \qquad (6.12)$$

$$\text{ConstraintNotification} := (id, \, type, \, sender, \, receiver, \, is \; answer, \, answer \; to,$$
$$ttl, \, distance, \, constrained \; message) \, , \qquad (6.13)$$

where *type* = 10, *is answer* = false, and *constrained message* $\in J$.

The constraint notification always replaces a demand notification or an offer notification. For routing purposes, the ID of the original message must be transmitted, for which the *constrained message* field exists.

In addition to the messages that make up the four-way handshake and, generally, deal with the flow of power, four other messages exist that primarily serve maintenance purposes.

An agent must send an *online notification* to notify its neighbors that it is becoming online, i.e., synchronized with the power grid, at a certain time. Likewise, an *offline notification* must indicate that the node an agent represents is going to be disconnected from the grid at a certain time. This allows agents to mark their connections to other agents as active or inactive, which is important for message routing.

$$\text{OnlineNotification} <: \text{Message} \, , \qquad (6.14)$$

$$\text{OfflineNotification} <: \text{Message} \, , \qquad (6.15)$$

---

[17]The 10-minute space is arbitrarily chosen, but based on the 10-minute interval used to calculate means in meteorological data. It assumes that a direct correlation exists between this fixed interval size and the message sending and processing behavior the multi-agent system exhibits.

$$\text{OnlineNotification} := (id, type, sender, receiver, is\ answer, answer\ to, \\ timestamp)\ , \tag{6.16}$$

$$\text{OfflineNotification} := (id, type, sender, receiver, is\ answer, answer\ to, \\ timestamp)\ . \tag{6.17}$$

For OnlineNotification, $type = 3$, and for OfflineNotification, $type = 4$. Both messages must not be answers, i.e., $is\ answer = $ false.

The protocol also implements the notion of *ping*, similar to the concept present in the *Internet Control Message Protocol* (ICMP), in the form of the *echo request* that must be answered by an *echo reply* if the node that receives the echo request is online with regards to its synchronization to the grid.

$$\text{EchoRequest} <: \text{Message}\ , \tag{6.18}$$

$$\text{EchoReply} <: \text{Message}\ , \tag{6.19}$$

$$\text{EchoRequest} := (id, type, sender, receiver, is\ answer, answer\ to, timestamp)\ , \tag{6.20}$$

$$\text{EchoReply} := (id, type, sender, receiver, is\ answer, answer\ to, timestamp)\ . \tag{6.21}$$

EchoRequest has $type = 1$ and $is\ answer = $ false, whereas EchoReply contains $type = 2$ and $is\ answer = $ false.

We can now see that messages are not only distinguished by their type and purpose, but also fall into three classes. Echo request, echo reply, and online and offline notification are link-local maintenance messages. We will later see that they are not forwarded; that they have no TTL field is an indicator of that. All other messages are eligible for routing. Additionally, two messages also carry power values: Demand notification and offer notification. Figure 6.3 illustrates these classes.

The protocol proposed in this thesis does not implement neighbor discovery; connections in the overlay network must be configured by another mechanism.[18] However, it implements service discovery: When an agent becomes connected to the communications network, it must send online notifications to its neighbors to indicate its node's synchronization to the power grid. This notifies the agent's neighbors that their connection is now online. In order for the agent coming

---

[18]This is primarily due to security reasons: See Section 6.4 for details on an agent's credibility.

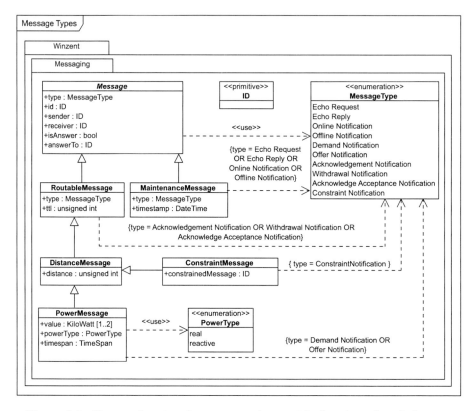

Figure 6.3: Types of protocol messages along with the class they belong to

online to know of the states of its own connections, it sends echo requests afterwards: Its neighbors must answer when they are online themselves. Hence, all agents have acquired the knowledge of their connections' states.

Message Processing Directives

No-Zombies Directive

The power grid knows circular structures and so does the overlay network created by this protocol. Thus, an agent may receive the same message several times. However, agents must also try to minimize the number of messages that travel the communication network at any given time. Especially 'dead' messages, i.e., those that are never answered, must be prevented from circulating endlessly.

Each message that must be routed by an agent contains a *ttl* field. The message's TTL is an integer without unit. Every time a message is to be forwarded by the agent, it must also decrement the message's TTL. If the value reaches 0, the message must not be forwarded. Instead, the agent must discard the message and also remove all data it has kept with regards to this message from its message journal.

A message's initial TTL obviously strongly influences the number of hops (i.e., routing agents) it can pass. An administrator must set it by policy, as the initial value is dependent on the size of the power grid.

Match-or-Forward Directive

A meshed network topology will cause a number of agents to receive demand notifications and offer notifications without a differentiation of whether they can actually fulfill the demand, either partially or completely. However, each agent must contribute to preserving the electrical grid's power equilibrium. Therefore, the Match-or-Forward Directive defines three behavioral rules an agent must follow upon the reception of a demand notification or offer notification:

1. If the agent cannot fulfill the request, but has previously sent a request of its own that would match the request received,[19] it must check whether to withdraw its own request or not:

   a) If the other request's *answer until* field indicates a time earlier than the agent's own request or the agent's own requested value is smaller than the other agent's requested value, it must withdraw its own request and issue an answer according to rules 3 and 4.

   b) Otherwise, the agent must continue with rules 2–4.

2. If the agent receives a constraint notification to a request it has previously answered and the constraint notification's distance value is equal to or lower than the distance value of the request the agent has answered, it must withdraw its response.

3. If the agent cannot fulfill the request, it must forward the message if no constraint prohibits this.

---

[19]E.g., the agent sent a demand notification and receives now an offer notification that is not an answer to its request.

4. If the agent can fulfill the request partially, it must

   a) formulate a response and send the response on the connection over which it had previously received the request, unless

      i. a constraint at the agent's node prohibits it, or
      ii. the agent has received a constraint notification with a distance equal or lower to the request received

   b) modify the request by the value it sent in the response, and forward the modified request.

5. If the agent can fulfill the request completely, it must not forward the message, but formulate an answer, unless

   a) a constraint at the agent's node prohibits it

   b) the agent has received a constraint notification with a distance equal or lower to the request received.

Through these rules, messages travel farther in the power grid only if they cannot be answered. This favors a local, distributed power generation, and most likely smaller power plants and nearby consumption over centralized generation, in which the power travels several kilometers of wire before reaching the consumer.

Forwarding Directive

The protocol creates an overlay network that models the power grid via the communications network and thus consumers and producers are not immediately connected with each other. Messages, such as demand notifications or offer notifications, are therefore routed by a number of agents representing, e.g., transformers. How agents should route or, in general, process messages they receive is governed by a set of directives every agent in must follow.

Forwarding denotes, the process of receiving a message and re-sending it. Not all messages are eligible for routing; some are only inteded to inform an agent's immediate neighbor. The messages that may be forwarded are:

1. Demand Notification

2. Offer Notification

3. Acceptance Notification

4. Withdrawal Notification

5. Acceptance Acknowledgement Notification

6. Constraint Notification.

In order to route messages efficiently, every agent must keep a data store of all messages received that are eligible for forwarding. This data store is called the *message journal*.[20] The message journal stores not only the message itself, but all of the links it was transmitted on, as well as the distance value for the message being transmitted on that particular link.

Definition 6.1. *Each agent's messaging module contains a message journal, denoted by the module-identifying letter M_i.*

The message journal contains a set of mappings:

$$
\begin{aligned}
M_i = \{ & m_1 \mapsto \{(l_{i,1}, m_{1,distance(l_{i,1})}), \ldots, (l_{i,n}, m'_{1,\text{distance}(l_{i,n})})\}, \ldots, \\
& m_n \mapsto \{(l_{i,1}, m_{n,\text{distance}(l_{i,1})}), \ldots, (l_{i,n}, m'_{n,\text{distance}(l_{i,n})})\}\} \ .
\end{aligned}
\tag{6.22}
$$

Each tuple contains the link, l_i, via which the message m was transmitted, as well as the transmission distance, of the respective variant of the message that was received via l_i, $m_{distance}$. The expression m' denotes such a variant of a message that has travelled via a different set of links than the first variant that was received, m.

The addition of a message with its link information to the message journal is expressed as:

$$
M_i \cup \{(m, l_{i,k})\} \ .
\tag{6.23}
$$

In order to retrieve the set of tuples a message $m \in M_i$ maps to, we write:

$$
M_{i,m} \ .
\tag{6.24}
$$

Since a message's ID identifies it unambiguously, it is indeed sufficient to use this ID value:

$$
M_{i,m_{ID}} \ .
\tag{6.25}
$$

---

[20]Prior publications (Veith et al., 2013, 2014) called this message store the "duplicate request cache," a name that was largely inspired by the idempotence of all messages, but that is now misleading since the cache is required for several tasks.

Hence, Eq. (6.24) and Eq. (6.25) can be used interchangeably.

The set of tuples is ordered through their distance. If a and b denote two tuples in the set $M_{i,m}$:

$$a \leq b \quad \Leftrightarrow \quad a_{distance} \leq b_{distance} . \tag{6.26}$$

Thus, the message journal stores, for each message, all connections over which it was received: an ordered set that is updated over time in meshed networks. The message journal also stores the smallest yet-encountered distance for this message (from Eq. (6.22): $M_{i,m,1}$). With regards to the message journal, the LPEP defines the equality between two messages according to a subset of their fields:

Definition 6.2. *In the LPEP, two messages m_1 and m_2 are equal, iff their IDs, sender, receiver, and type fields are equal. Formally, the equality operator with regards to messages is defined as:*

$$
\begin{aligned}
m_1 = m_2 \quad \Leftrightarrow \quad & m_{1,ID} = m_{2,ID} \\
& \wedge\, m_{1,type} = m_{2,type} \\
& \wedge\, m_{1,sender} = m_{2,sender} \\
& \wedge\, m_{1,receiver} = m_{2,receiver} .
\end{aligned}
\tag{6.27}
$$

Therefore, two variants of a message, m_1 and m'_1 are equal, even fields not listed in Eq. (6.27) are not equal. Thus, the equality operator does not express the identity of two messages.

Upon reception of a routable message, the agent must first store it in its message journal. Then, it must check whether it is the designated recipient or not. If the agent is indeed the intended recipient, it must act on it. Otherwise, the agent acts as a router and may forward the message, according to the rules laid down in Section 6.2.

In order to select the best connection to forward a message over, agents require a metric, in a way similar to the hop count of the IP. Hence, each message contains a *distance* field. Since the LPEP models the power grid, this distance metric is based on the actual power flow at each node. The origin agent initializes the distance field as 0; every forwarding agent then adds the distance of the connection it chooses: This distance is the impedance of the power line the link models.

Definition 6.3. *The set of links, $L_i, i \in I, l_{i,k} \in L_i, k = 1, 2, \ldots, |L_i|$, is an ordered set of individual links or connections for each agent, A_i. The set is ordered by the impedance of each line at the time the contents of the message for which the current ordering is retrieved will take effect[21], $Z_{i,k}(t)$. Thus, the distance of each connection is:*

$$l_{i,k,\text{distance}}(t) = Z_{i,k}(t) \tag{6.28}$$

Furthermore, the ordering of links in L_i is expressed through the smaller-than-or-equal operator, $\leq$:

$$l_{i,1}(t) \leq l_{i,2}(t) \quad \Leftrightarrow \quad l_{i,1,\text{distance}}(t) \leq l_{i,2,\text{distance}}(t) \tag{6.29}$$

For re-sending the message, the agent selects $[0; |L| - 1]$ connections according to these rules:[22]

1. If the message type is one of echo request, online notification, or offline notification, it must not be forwarded and the agent must process it locally.

2. If the agent is the designated recipient, it must process the message locally.

3. If the message is an answer, the agent is not the indicated recipient, and $m_{ttl} \geq 1$ is true, the agent must decrement the message's TTL and afterwards check for the message in its message journal:

   a) If the message is contained in the agent's message journal, it must forward it over the connection with the lowest distance available, i.e., $\text{argmin}_{l_i \in M_{i,m}} \text{distance}(l_i)$.

   b) If the message is not contained in the agent's message journal, it must forward the message over all connections but the receiving one.

4. If the message is not an answer and $m_{ttl} \geq 1$ is true, the agent must decrement the message's TTL. It must then apply the Match-or-Forward Directive. If the processing of the message according to the directive indicates that the message must still be forwarded, it must select $[0; |L| - 1]$ connections to forward the message over:

---

[21]Cf. the definition of messages indicating a change in power flow in Eqs. (6.4) and (6.5).
[22]These rules also include the No-Zombies Directive and the Match-or-Forward Directive.

a) If the agent has no record of the message in its message journal, it must create a record for it and must then select the links on which it will forward the message:

  i. The agent must not forward the message over the receiving connection.

  ii. If a constraint exists for the node the agent represents or the candidate link the message could be forwarded on that prohibits forwarding of the message, the agent must replace the message with a constraint notification and forward the new constraint notification instead.

  iii. If power is allowed to flow through the node the agent represents and via the respective candidate link, the agent must forward the message on the link.

b) If the agent has a record of the message, but the message's distance value is smaller than the distance recorded in the agent's message journal, it must forward it over all connections over which the message was not previously received, except if prohibited by a constraint (see above).

c) If the agent has a record of the message and the distance value of the now received variant of the message is greater than or equal to the recorded distance value, the agent must not forward it.

Through Definition 6.3, an agent will choose the connection with the least distance as candidate link for routing[23] a certain response. It will choose the connection that represents that power line over which the requested power will eventually flow.

Once an agent receives a message that is an answer, it may, after forwarding it, remove all records of the previous request. It may do so likewise once the time indicated in the *answer until* field has passed.

6.3 Data Encoding

JSON Encoding

The LPEP is a message-based application-level protocol[24] and therefore initially free to choose any encoding. With regards to embedded devices, where

---

[23]Meaning, it must have received the matching non-answer request on the link.
[24]Cf. Section 6.1.

computing time is precious and the data rate of the device's connection is low, an efficient format is desirable. While this suggests a binary encoding, it is important not to hinder debugability. The LPEP therefore offers both formats: the JSON as human-readable, yet efficiently-to-parse, format and a binary encoding. This section concerns itself with the description of the JSON encoding, while Section 6.3, which follows, specifies the binary encoding.

Every message of the LPEP is a JSON object; the aforementioned keys in Eqs. (6.1) and (6.21) are converted to 'mixedCase' notation. Abbreviations, such as TTL, are converted to lower case. Thus, the keys in Eq. (6.1) are `id`, `type`, `sender`, `receiver`, `isAnswer`, and `answerTo`.

Since all identifiers are opaque, there is no need to superimpose a particular structure. However, they must be represented by ASCII characters to both follow the JSON standard[25] that allows unicode notation and to allow humans to distinguish between different identifiers.

For encoding timestamps, the *Temps Atomique International (en. International Atomic Time)* (TAI) standard is used: More specifically, timestamps are represented by TAI64 labels (Bernstein, 1997). Textual encoding represents the label as a string of hexadecimal characters. Thus, the TAI64 external format is suitable for textual as well as binary encoding. For example, the timestamp `2015-12-08 13:31:54 +0000` would be represented in TAI64 external format as `400000005666dbed`.

Time intervals are an array of two elements, denoting the two boundaries of the interval. With regards to Eqs. (6.4) and (6.5), the LPEP only offers one type of time interval.

Power values are represented as integers, the power type (`powerType` key) as a string of either `active` or `reactive`.

E.g., a demand for (exactly) 5000 kW in the time interval 2015-12-01 00:00:00 +00:00 to 2015-12-01 01:00:00 +00:00 could[26] be encoded in the LPEP JSON format as:

---

[25]Cf. (Bray, 2014).

[26]Because identifiers are opaque, the subjunctive is appropriate here.

```
{
  "id": "c40c67e2-477d-45fe-b29c-84ea134d5d97",
  "type": 5,
  "sender": "bareHillWindfarm",
  "receiver": null,
  "isAnswer": false,
  "answerTo": null,
  "ttl": 42,
  "distance": 0,
  "timespan": ["40000000565ce323", "40000000565cf133"],
  "answerUntil": "40000000565cdf9f",
  "value": [5000, 5000],
  "powerType": "active"
}
```

Binary Encoding

The binary encoding format of the LPEP offers fixed-width fields for fast processing of messages. For the same reason, all fields are multiples of 16 bits wide. Each 16 bit group byte order is the *network byte order*, i.e., *big endian*.[27] The order of the fields is given through the message type definitions in Eqs. (6.1) and (6.21). The message base type is encoded as shown in the schema in Table 6.1.

Since all fields are uniformly 16 bits wide, Boolean values will also take up that width. The *type* field is therefore a 16 bit-wide enumeration; a Boolean `true` is encoded as 16 1-bits, a Boolean `false` as 16 0-bits. A timestamp is, through the TAI64 label, 8 bits wide; a timespan, i.e., a closed interval consisting of two TAI64 labels, is a concatenation of the upper and lower boundary and therefore 16 bits wide.

All values $x \in \mathbb{N}$, i.e., the *ttl*, *distance*, and the acceptance notification's *value* fields are encoded as 32 bit-wide unsigned integers. The power value interval a demand or offer notification carries is a concatenation of two power values, hence 64 bits wide. Finally, the *power type* field is encoded as an enumeration similar to the *type* field, with active $\mapsto 0$ and reactive $\mapsto 1$.

A complete reference of all field types, along with their encodings, is given in Appendix B.2.

---

[27]Meaning that the most significant bit is the first to be transmitted

Table 6.1: Binary encoding schema of the Lightweight Power Exchange Protocol message header

0–15	16–31	32–47	48–63	64–79	80–95	96–111	112–127
id							
type	sender...						
...snd.	receiver...						
...recv.	is answer	answer to...					
...answer to		...					

6.4 Analysis

Implementability

One can assume that the basic facilities to implement an application-layer protocol exist in most, even embedded, devices: A network stack and possibly a JSON parser. The implementability of the protocol therefore depends on an existing (pseudo-) code description that shows the transcription of the rules formulated in natural language in Section 6.2 into a concise format.

At first, we can divide the message processing between two main groups of messages: Those that are eligible for forwarding, which are all of the main message types necessary for the distributed demand-supply calculation, and the maintenance methods echo request and echo reply, as well as the connection status update messages, i.e., online notification and offline notification. The initial ProcessMessage function therefore distinguishes between these groups.

Algorithm 3 Lightweight Power Exchange Protocol message processing

---

procedure PROCESSMESSAGE(m, l)
    global L_i : Links, M_i : MessageJournal
    if $m_{type} \in$ {EchoRequest, EchoReply,
           OnlineNotification, OfflineNotification} **then**
        PROCESSMAINTENANCEMESSAGE(m, l)
    else if $m_{type} \in$ {DemandNotification, OfferNotification,
           AcceptanceNotification, AcceptanceAcknowledgementNotification,
           WidthdrawalNotification, ConstraintNotification} **then**
        PROCESSPOWERMESSAGE(m, l)
        $M_i \leftarrow M_i \cup (m, l)$ ▷ New Message Journal record (cf. Eq. (6.23)).
    else
        return   ▷ Discard bogus message.
    end if
end procedure

The last mentioned maintenance message types are exclusively used to update status information about the agent's immediate neighborhood:

Algorithm 4 Handling of maintenance messages

---

procedure PROCESSMAINTENANCEMESSAGE(m, l)
    global L_i : Links
    if m_{type} = EchoRequest **then**
        SEND(r : EchoReply(m))
    end if
    if m_{type} = EchoReply **then**
        UPDATESTATUS(m)
    end if
    if m_{type} = OnlineNotification **then**
        for all $l \in \{L_i \mid m_{sender} = l_{destination}\}$ **do** $l_{status} \leftarrow$ online
    end if
    if m_{type} = OfflineNotification **then**
        for all $l \in \{L_i \mid m_{sender} = l_{destination}\}$ **do** $l_{status} \leftarrow$ offline
    end if
end procedure

Program logic regarding power messages, i.e., those that are the agents' vocabulary regarding the distributed planning process, are handled either locally or remotely: Revisiting the Match-or-Forward and Forwarding Directive outlined in Section 6.2, we see that each agent does a portion of the planning process, while the other part happens remotely, i.e., through forwarding a modified version of the message. The program logic regarding local evaluation is handled in the `ProcessPowerMessage` function:

Algorithm 5 Processing of power messages

---

procedure PROCESSPOWERMESSAGE(m, l)
    global A_i : Governor, L_i : Links, M_i : MessageJournal, P_i : PowerBalance
    if $type_m \in \{$DemandNotification, OfferNotification$\}$ **then**
        f_m : Forecast $\leftarrow m$ ▷ Convert message contents to Forecast
        if $\neg m_{isAnswer}$ **then**   ▷ Request (broadcasted)
            if HASCONSTRAINT(A_i, m)$\vee$CONSTRAINTRECEIVED(M_i, m) **then**
                $constraintMessage \leftarrow$ ConstraintMessage(m)
                FORWARD($constraintMessage$)
            else if $\{-f_m\} \cap P_i \neq \emptyset$ **then**   ▷ Resolve 'clash of requests.'
                $request$: Message $\leftarrow$ GETMESSAGETOFORECAST($-f_m$)
                if $answerUntil_m < answerUntil_{request}$
                        $\vee value_{request} < value_m$ **then**
                    $M_i \leftarrow M_i \cap request$
                    for all $l \in L_i$ **do**
                        SEND(l, r : WithdrawalNotification($request$))
                    end for
                    $response$: Message
                    if $m_{type} =$ DemandNotification **then**
                        $response \leftarrow$ OfferNotification($P_i \cap \{-f_m\}$)
                    else
                        $response \leftarrow$ DemandNotification($P_i \cap \{-f_m\}$)
                    end if
                    $M_i \leftarrow M_i \cup response$
                    SEND($l, response$)
                end if
            else if $\{f_m\} \cap P_i \neq \emptyset$ **then**   ▷ Check for possible fulfilment.
                $response$: Message
                if $m_{type} =$ DemandNotification **then**
                  $response \leftarrow$ OfferNotification($P_i \cap \{f_m\}$)
                else
                  $response \leftarrow$ DemandNotification($P_i \cap \{f_m\}$)
                end if

$\qquad M_i \leftarrow M_i \cup response$

$\qquad \textsc{Send}(l, response)$

$\quad$ **end if**

else $\qquad\qquad\qquad\qquad\qquad\qquad$ ▷ Message is a reply...

$\quad$ **if** $m_{destination} = i$ **then** $\qquad$ ▷ ...to a request of our own.

$\qquad P_i \leftarrow P_i \cup \{f_m\}$ $\qquad\qquad$ ▷ Triggers search for solution.

$\quad$ **else** $\qquad\qquad\qquad\qquad\qquad$ ▷ ...to another agent's request.

$\qquad \textsc{Forward}(m)$

$\quad$ **end if**

end if

else if $m_{type} = \text{AcceptanceNotification} \wedge M_{i,m} = \emptyset$ **then**

$\quad response \leftarrow \text{AcceptanceAcknowledgedNotification}(m)$

$\quad \textsc{Send}(l, response)$

else if $m_{type} = \text{WithdrawalNotification}$ **then**

$\quad f_m : \text{Forecast} \leftarrow m$

$\quad P_i \leftarrow P_i \setminus \{f_m\}$

else if $m_{type} = \text{AcceptanceAcknowledgementNotification}$ **then**

$\quad \textsc{FinalizeHandshake}(m)$

else if $m_{type} = \text{ConstraintNotification}$ **then**

$\quad record \leftarrow M_{i,m_{constrainedMessage}}$

$\quad$ **if** $record \neq \emptyset$ **then**

$\qquad response : \text{Message} \leftarrow \textsc{GetResponse}(record_{1,message})$

$\qquad$ **if** $response \neq \emptyset$ **then**

$\qquad\quad withdrawal : \text{Message} \leftarrow \text{WithdrawalMessage}(response)$

$\qquad\quad \textsc{Send}(l, withdrawal)$

$\qquad$ **end if**

$\quad$ **end if**

$\quad \textsc{Forward}(m)$

end if

$M_i \leftarrow M_i \cup (m, l)$ ▷ Create new Message Journal record (cf. Eq. (6.23)).

end procedure

Finally, forwarding of—processed—messages is the duty of the `Forward` function, whose main task is the selection of the proper outgoing links. It takes care of the TTL management and discards zombie messages.

Algorithm 6 Forwarding of Lightweight Power Exchange Protocol messages

---

procedure FORWARD(m)
    global M_i : MessageJournal, L_i : Links
    if $m_{type} < 5$ **then**
        return   ▷ Only certain message types may be forwarded.
    end if
    $m_{ttl} \leftarrow m_{ttl} - 1$
    if $m_{ttl} \leq 0$ **then**
        return   ▷ Discard zombie messages.
    end if
    $record \leftarrow M_{i,m}$ ▷ Per Eq. (6.24)
    if $m_{isAnswer}$ **then**
        $bestConnection \leftarrow record_{1,connection}$
        $m_{distance} \leftarrow m_{distance} + bestConnection_{distance}$
        SEND($bestConnection, m$)
    else
        for all $l \in L_i$ **do**
            $m_{distance} \leftarrow m_{distance} + l_{distance}$
            if $l \notin record$ **then**   ▷ Forward on not already used links.
                if $m_{type} \in \{$DemandNotification, OfferNotification$\}$
                      $\wedge$HASCONSTRAINT(l, m) **then**
                    $constraintMessage \leftarrow$ ConstraintNotification(m)
                    SEND($l, constraintMessage$)
                else
                    SEND(l, m)
                end if
            else if $distance < record_{1,distance}$ **then**
                SEND(l, m)   ▷ Update distance value for others.
            end if
        end for
    end if
end procedure

Scalability

The Match-or-Forward and Forwarding Directives, defined in Section 6.2, divide the behavior of LPEP agents managing demand and supply into two distinct parts:

1. A broadcasting query or advertisement stage

2. a unicast response or direct-addressing stage.

These two stages of a demand-supply communication are easily found in the messages themselves, namely in the *is answer* indicator. A message that is eligible for forwarding and is not an answer is broadcasted; any answer is ideally relayed on a direct route.[28] This direct routing is enabled through the agents' message journal, in which each forwarded message must be recorded.[29]

During broadcasting, messages are essentially duplicated several times. Every k-th agent via which such a demand or offer notification travels has a set of links, L, and selects at most $|L| - 1$ connections. This upper boundary equals the total number of connections an agent maintains minus the receiving link iff the message travels a network without loops during its TTL, which requires a tree-like network structure.

On a network containing loops, the agent must not select links over which the message has previously travelled for sending, as per the Forwarding Directive. Therefore, every k-th agent looks up the number of connections already used for delivering the message in its message journal, $M_{k,m}$. Forming the complement with the set of all links on the k-th agent, L_k, we retrieve all candidate links with the expression $L_k \setminus M_{k,m}$. This also includes the receiving connection, since the message journal stores both incoming and outgoing versions of the same message. Thus, the number of connections used in the broadcast stage, $|L'|$, is:

$$|L'| = \sum_{k=1}^{m_{ttl}} \left(|L_k| - |L_k \setminus \{l : l \in M_{m,k}\}| \right) \ . \tag{6.30}$$

While Eq. (6.30) shows the average-case complexity of the LPEP's broadcasting stage, the worst-case complexity is attained in purely meshed structures when distance and delay values of the links are distributed in inverse proportion, i.e., when links with low distance values have high transmission delays and vice versa. Such a mesh can be easily constructed with any number of nodes.

---

[28]This assumes that a broadcasted request precedes any answer. This is true except when the network topology changes between a request (i.e., a demand notification or an offer notification) and the response. If no corresponding record is found, the message must be broadcasted, too. Cf. the Forwarding Directive in Section 6.2 for details.

[29]A message may only be removed from the message journal if the *answer until* timer of the original request message has been expired.

Through the requirement to send distance updates, the total number of links that will see a variant of the message will be:

$$|L'| = |L_0| + (|I| - 1) \cdot |L \setminus L_0| \, . \tag{6.31}$$

Here, L_0 represents the set of links of the initiating agent; over these links no distance update will ever be transmitted. The rest of the term represents all other agents. Assuming that the dominating term is the one that represents the forwarding action of all non-initiating nodes, we can rewrite the term as:

$$|L'| \approx 2 \cdot |I| \cdot (|L| - 1) \, . \tag{6.32}$$

Since we assume a meshed structure, all $|I|$ nodes then have $|L_i| = |I| - 1$ connections. We therefore arrive at the worst-case complexity of:

$$|L'| \approx 2 \cdot |I|^2 - 4 \cdot |I|, \quad \text{given } |I| \geq 3 \, , \tag{6.33}$$

$$\mathcal{O}\left(|I|^2\right) \, . \tag{6.34}$$

This is normal for an algorithm based on broadcasting and also accepted behavior of other protocols, e.g., link-state routing protocols that continuously update their link state databases through flooding.[30] Since no a priori knowledge of potential partners is available and no global view of the network exists so that the search can stop early—such as a tree search might—, the worst-case complexity in Eq. (6.33) remains. However, connections between whole branches that create loops at the boundaries of the network are atypical for the power grid. Therefore, the worst case complexity will remain theoretical.

The decentralized approach excludes other, only seemingly more efficient, alternatives for two reasons:

1. The state of each node is encapsulated by the corresponding Agent instance and there is—by purpose—no global knowledge about each node's forecasts. Even if a sink tree was available, without additional information, a requesting node would still have to address all other known nodes as no knowledge about the other nodes' states can be immediately available.

---

[30]Cf., e.g., OSPF (Coltun et al., 2008), where updates to the *Link State Database* (LSD) are propagated through flooding and the information of reachability for a certain subnet therefore travels in a similar fashion.

2. Routing protocols such as OSPF that make use of sink trees also take advantage of aggregations, such as those offered by IP's subnets and the different area definitions in OSPF[31] that are not available for the power grid: Whether a certain power subnet would be a consumer or producer at a certain point depends on the communication of forecasts and forecasting patterns, which change over time, necessarily leading to more communication at another point in the protocol design.

A centralized approach can be excluded for obvious reasons—the same that have been discussed in Section 4.1. We can thus conclude that the worst-case complexity of the LPEP is without an alternative.

The second stage of the LPEP communication behavior encompasses the routing of answers. Here, we can assume that the path of the answer is contained in the set of subgraphs through which the original request travelled. The optimal route, from the perspective of the power grid, is contained as distributed information in the message journals of the agents forming the path of the answer. Since, according to the Forwarding Directive,[32] an agent must select the best outgoing link when forwarding, the complexity of the answer is linear and equal to the difference of the requests original TTL (ttl) and the remaining TTL count for the lowest distance path ($ttl' = ttl_{\min(m_{distance})}$):

$$\mathcal{O}\left(request_{ttl} - request_{ttl'}\right) \ . \tag{6.35}$$

We must now further examine the protocol's scalability by comparing an existing network to a possible extension of it. Consider Fig. 6.4. When we assume that all existing connections are ideal (i.e., have no latency and no data loss), we immediately see that agents roughly fall into two categories: Router nodes and end points. Router nodes have more than one connection ($|L| > 1$), whereas an endpoint only has one. Although every connection is bidirectional,[33] upon forwarding, the receiving connection may never be used to forward the same[34] message. Router nodes select at most $|L| - 1$ connections. If we further assume in this simple scenario that all connections share the same metric, no request message passes the same node twice.

An extension of the network can happen in two ways: Either by creating a star-like topology, effectively forming a sub-graph beginning with the router node.

---

[31] Since less and less subnet summarizing is happening in the IPv4, routing tables are still growing, leading to issues described by Atkinson (1996); Santos (2014); Sverdlik (2014).

[32] Cf. Section 6.2.

[33] Cf. the connection concept in Section 6.2.

[34] I.e., $m_1 = m_2$ as per Definition 6.2

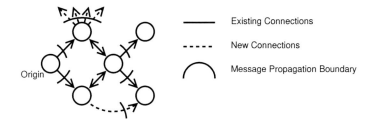

Figure 6.4: Sample network with message propagation boundaries

The agent node, which the subgraph is connected to, is then the natural message propagation boundary, as per the Forwarding Directive. A new connection may also create a loop, but even then does a router node pose a message propagation boundary. Hence, the worst-case described in Eq. (6.33) assumes that the request is not answered and that the message is able to travel through various non-circular sub graphs, since loops do not lead to message duplication as per the Forwarding Directive and Definition 6.2.

This simple scenario deliberately considered only ideal circumstances. It left out a rule of the protocol: An updated forwarding of a request that happens if a router node receives the same message twice, but the second time with a lower distance value.[35] This happens especially in conjunction with different delays on the data links that are used for the corresponding agent connections: The power line the connection represents may have a low distance value, but uses a high-delay data link, whereas the high-distance connection benefits from a low-delay data link. In this case, delay and distance of the two connections are anti-proportional properties.

Let us therefore return to the reference grid described in Section 3.2 and consider, for actual simulation runs, the sub-grid created by the 'White Hill Springs Substation.' This network offers tree-like structures (e.g., starting from the 'White Hill Springs' transformer #1) as well as circular ones. Consider a request—more specifically, a demand notification—originating at the 'Bare Hill Wind Farm.' The network can be configured to create the best case, the average case, and the worst case.

This distinction is based on the proportionality (or anti-proportionality) of the two properties, delay and distance, of each link between the agents. These two properties stem from the two worlds the LPEP bridges. The data

---

[35]Cf. the Forwarding Directive.

network aspect introduces the transmission delay of the links within the overlay
network, whereas the power grid aspect introduces the distance as defined in
Definition 6.3.

In the best case, the delay and the distance are proportional to each other:
A link with a low distance will transmit a message faster than a link with a
high distance, favoring the semantics of the protocol because no single node is
required to send distance update packets. In the worst case, these two properties
are anti-proportional to each other, creating distance updates for each set of
parallel links. For the average case, typical values for impedance and delay are
randomly chosen during repeated runs. Here, wind farms are considered to be
connected using UMTS or a similar wireless technology (Chan and Ramjee,
2005), whereas the transformers use typical wire technology, up to as good as
standard consumer connectivity, such as *Asymmetric Digital Subscriber Line*
(ADSL).

In order to observe the message propagation during the request/broadcasting
phase, we inject a local disequilibrium at the 'Bare Hill Wind Farm' site and
define the initial simulation state, S_0, to contain the power balance following
the syntax defined in Eq. (3.11):[36]

$$P_{bareHillWindFarm,0} = \{$$
$$([2014\text{-}02\text{-}19\text{T}12\text{:}00\text{:}00\text{+}01\text{:}00; 2014\text{-}02\text{-}19\text{T}12\text{:}10\text{:}00\text{+}01\text{:}00), 760\,\text{kW}) \} \ .$$

$$(6.36)$$

When observing the message propagation, circular and tree-like structures
influence the number of copies of a message that are being transmitted in
different ways.

Purely circular structures such as the cutout of 'Saltwater Town' (SWT)
shown in Fig. 6.5 always create at least $|I'| + 1$ messages, where I' is the set of
nodes in the (sub-) structure. At most, in the worst case, $2 \cdot |I'| + 1$ messages
are sent. This happens when all links in the subset except for one entry link
have low delay values and all links except for the second entry link have low
distance values. In Fig. 6.5, these two distinct links are those between 'SWT
Trafo #1,' and 'SWT Trafo #2' and 'SWT Trafo #3' respectively.

Tree-like structures never generate distance updates, only loops do. However,
all structures must forward distance updates until they reach the end points

---

[36]The propagated value is the difference between the power balance at 11:50 and 12:00,
which is valid for the corresponding interval [11:50; 12:00) and [12:00; 12:10).

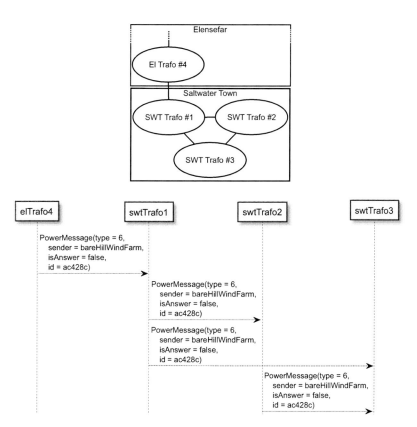

Figure 6.5: Message propagation in the 'Saltwater Town' part of the reference grid

of the grid, which act as natural message boundaries in the same way as any sub-structure does, as Fig. 6.4 indicates.

Table 6.2 lists the results of the best, the average, and the worst case simulation runs for the 'White Hill Springs Substation' subgrid. Note that the numbers for the average case are rounded to integers since 'half a message' does not exist: A message is either sent or not sent.

In order to observe the propagation of a response, we enable the block and heat power plant of the reference grid to answer the request of the 'Bare Hill Wind Farm' by ramping up power output. Additionally, we fix the TTL of new messages to known value:

Table 6.2: Messages forwarded in the reference grid during the broadcasting stage

Agent Name (i)	$|L_i|$	$\min(|M_i|)$	$\mathrm{avg}(|M_i|)$	$\max(|M_i|)$
Fool's Springs Wind Farm	1	0	0	0
Block and Heat Power Plant	1	0	0	0
Ws Trafo #1	2	1	3	3
Wh Trafo #1	3	2	6	6
White Hill Wind Farm	1	0	0	0
Blows Hill Wind Farm	1	0	0	0
White Hill Springs Substation	5	4	12	12
Lambert Springs Wind Farm	1	0	0	0
Fd Trafo #1	2	1	3	3
Fl Trafo #1	3	2	4	5
Fl Trafo #4	2	1	2	3
Fl Trafo #5	3	2	4	4
Fl Trafo #7	4	3	6	6
Fl Trafo #11	2	1	3	3
Augustus Works	1	0	0	0
Fb Trafo #1	3	1	1	1
Bw Trafo #1	2	1	1	2
Bw Trafo #6	3	2	2	2
Bare Hill Wind Farm	1	1	1	1
El Trafo #1	4	3	5	6
El Trafo #4	2	1	2	4
El Trafo #6	2	1	1	2
El Trafo #8	3	2	2	4
El Trafo #12	3	2	4	7
El Trafo #17	2	1	2	2
Levee's Pillow Factory	1	0	0	0
SWT Trafo #1	3	2	4	8
SWT Trafo #2	2	1	2	5
SWT Trafo #3	2	1	2	4
Total	32	36	72	93

$$ttl_0 = 42 \ . \tag{6.37}$$

This way, we can formulate the expected states of the message journals of those agents who will forward the message on the direct route, both for the best and the worst case scenario. For example, 'Funder's Village Trafo #1' will forward the message in both cases directly and the final and desired state will be:

$$M_{fdTrafo1,T} = \{(type = 7, sender = Block\ Heat\ and\ Power\ Plant, ttl = 38)\} \ . \tag{6.38}$$

Knowing the distance values beforehand, we can formulate final state definitions for every agent along the way. Most agents will not route the message and due to Eq. (3.13) we cannot formulate an exclusive final state set member, i.e., we cannot formulate that an agent must not have received and forwarded the response. However, since we know the initial TTL, we can check that, in every case, the 'Bare Hill Wind Farm' agent has received the response on the shortest route, considering that the message distance metric also indicates the shortest route in terms of edges being that with the lowest distance:

$$M_{fdTrafo1,T} = \{(type = 7, sender = Block\ Heat\ and\ Power\ Plant, ttl = 32)\} \ . \tag{6.39}$$

Adding a new agent to the path of a response message will add one more link and therefore one more hop to it. The number of links the message travels is therefore:

$$|L''| = |L'| + 1 < |L'|^2 \ . \tag{6.40}$$

During the broadcasting phase, the message will travel over every link, potentially multiple times. We can therefore consider the deliberately introduced worst case for the request's propagation and compare it to the best case. On average, the number of messages sent corresponds to Eq. (6.30). Since, upon the addition of a new agent to the grid, the average complexity does not rise polynomially, i.e.,

$$|L'| + |L_{i'}| < |L'|^2 \ , \tag{6.41}$$

during the broadcasting phase for requests, the LPEP scales reasonably well under the given constraints.

Security Issues

After we have examined the scalability properties of the LPEP, we need to turn to securing the agents' communication. Due to the distributed nature of the approach presented in this work, network security may very quickly prove to be its Achilles' heel.

Two major points of attack present itself: A compromise within the protocol, which leads to false behavior or outages, or a compromise of an Agent instance.

The LPEP creates an overlay network, as discussed in Section 6.2, over the existing communication infrastructure. This makes it potentially vulnerable to any attacker or agency that is able to intercept data being sent between the agents. It is therefore prudent to secure the overlay network with an appropriate technology, such as a *Virtual Private Network* (VPN).

But even then an attacker could exploit a weakness in the system, since a VPN can be successfully attacked as well as any other system. Provided the attacker has gained access to the LPEP overlay network, he can compromise the agent's communication in a number of ways.

As a man-in-the-middle, he can intercept messages and alter them. There are obvious ways to cause damage this way by, e.g., altering the *value* field in any demand notification or offer notification. The modification of the message's *distance* field would be a more subtle attack, whereby the attacker could force an agent to prefer particular nodes over others.

Instead of modifying existing messages, the attacker could also use agent IDs he has learned by sniffing to forge complete messages, thereby injecting faux data into the network. Forged offline notifications can disable working connections and false requests initiate a bogus planning stage. While responses to an unsent request will be ignored by the agent that allegedly initiated the exchange, it can bind resources that would be denied to other, serious requests: The responder will truthfully send an answer with the latest possible value in the *answer until* field and must not respond twice to the same request, or two requests in the same time interval. The response will become invalid automatically after the deadline has passed, but until then, real requests will remain unanswered.

A man-in-the-middle will also be able to attack a specific node in a more profane manner by simply overloading it. Well-known *Denial of Service* (DoS) attacks attempt to flood a system with possibly bogus messages so that it is unable to answer serious communication attempts.

Such a DoS attack can also be performed from the outside: Since the LPEP forms an overlay network on the application level in the ISO/OSI stack model,

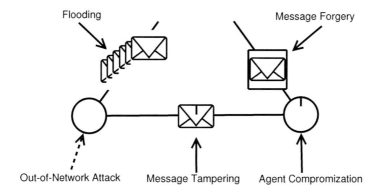

Figure 6.6: Possible attack vectors

its nodes must communicate on the network stack's network layer like any other host on the Internet. Obviously, this makes an agent vulnerable to all typically known forms of attack any server on the Internet faces. However, the LPEP's architecture can provide a way to mitigate these attacks: The agent's partners are known since all connections in the LPEP are bi-directional, point-to-point connections. Thus, the public (IP) addresses of the connected agents are also known and a firewall configuration to limit the set of allowed partners for incoming transmissions can be produced by an administrator. The operating system the agent runs on can simply drop unwanted incoming packets, an option not available to the typical Internet server where each connection attempt can be initiated in earnest. Furthermore, DoS and especially *Distributed Denial of Service* (DDoS) attacks are hard to distinguish from the activity of regular users.

Of course, physical access to the machines themselves must be protected, because once an attacker has gained access to the actual hardware, his possibilities for intrusion and manipulation are plentiful. If this can be ruled out, a node can only be compromised from the outside, by either exploiting a security hole in publicly available services on the node or by compromising a software update.

Verifying software updates is a common practice in all current operating systems and is typically achieved by signing software packages with a trusted key. Of course, this key must never be lost, since it must then be revoked by a central authority.

The concept of a trusted third party is known to be problematic and has become so specifically in the context of certificates for websites (i.e., HTTPS connections). First, the trusted third party is usually ultimately trusted, but opaque to users, even to those who receive certificates issued by the *Certificate Authority* (CA).[37] Even if the CA has been compromised, or changes its behavior to a malevolent one during its lifetime, certificates issued by it are still trusted: A problem that has already had impact in practice (VASCO Data Security International, Inc., 2011; Nationaal Cyber Security Centrum, 2011; Coates, 2013).

Even if the centralized CA does not per se cause any problem, certificate revocation is also highly problematic. The *Online Certificate Status Protocol* (OCSP) has been proven it be defeatable (Marlinspike, 2009), rendering the idea of a centralized, trusted third party mostly useless.

A decentralized contrast known in cryptography is the *Web of Trust* (WoT). Here, different users sign each other's public keys after they have verified the authenticity of the corresponding key-owner relationship (Ferguson and Schneier, 2003, p. 333). The idea of the WoT has been applied to peer-to-peer systems (Wang and Vassileva, 2003; Xiong and Liu, 2003), and can be applicable to the distributed architecture of the LPEP.

The idea of the WoT includes the notion that any decrease of trust in an agent can only happen after some incident, as other parties are revoking their own statements of trust for the particular node. In order to prevent an incident beforehand, malicious modifications of the agent software must be prevented. Kernel-level auditing systems, such as SELinux (Jaeger et al., 2003; National Security Agency, 2013), can identify misbehaving programs and halt their execution by disallowing certain system calls before any damage is done.

In the end, this section can only give pointers as to where security issues are present and what possible approaches may solve them. Since the author is not proficient with internet security, this section cannot and must not serve as a complete analysis of attack vectors and their mitigation: Worse than no security is poor security, as it gives a false sense of security and allows an able attacker easily to overcome low hurdles, whereas a lulled administrator will not re-evaluate his security concept on a regular basis.

---

[37]The CA is said trusted third party.

7 Modeling and Calculating Demand and Supply for Agents

7.1 Agent-Local Power Balance

The Universal Agent models its environment in two forms: Its neighborhood that is constituted of its fellow agents it communicates with, along with their messages, and the current state and future state of its own locality, the power balance.

Each agent keeps account of the power generation and consumption of its local node. It can be very much seen as a 'power ledger.' However, the goal of the agent is to maintain a power equilibrium at all times and insofar as the term 'power ledger' is similar to the financial term, 'balance' is more appropriately used here. The agent's power balance stores mappings of time intervals to power values, i.e.,[1]

$$[t_1; t_2) \mapsto P \ . \tag{7.1}$$

With regards to the desired equilibrium, an agent must fundamentally consider each entry in the power balance as a *requirement*: Each mapping individually constitutes a power imbalance, i.e., a disequilibrium, and therefore requires the agent to act. The power balance is indeed then balanced when individual requirements are matched in such a way that they equalize each other; thus demand and supply is counterbalanced. The power balance is, therefore, the keystone of the agent's fundamental goal.

---

[1]All equations presented in this chapter use the symbol P for denoting power. In a strict sense, this only means active power. However, the type of power is transparent for the equations and algorithms presented here. I.e., an analogous variant for reactive power also exists in parallel, but is not explicitly mentioned.

Internally, the agent represents the power balance as an *interval map*, extending the notion of an *interval set* with mappings, using the relevant mathematical operations on sets as well as defining operations on interval overlap:

$$\{[t_1; t_2) \mapsto P_1\} \cup \{[t_1; t_2) \mapsto P_2\} = \{[t_1; t_2) \mapsto P_1 + P_2\} \,, \qquad (7.2)$$

$$\{[t_1; t_2) \mapsto P_1\} \setminus \{[t_1; t_2) \mapsto P_2\} = \{[t_1; t_2) \mapsto P_1 - P_2\} \,. \qquad (7.3)$$

Partial overlaps are calculated analogously. Intervals of four times, $t_1 < t_2 < t_3 < t_4$, result in:

$$\{[t_1; t_3) \mapsto P_1\} \cup \{[t_2; t_4) \mapsto P_2\}$$
$$= \{[t_1; t_2) \mapsto P_1, [t_2; t_3) \mapsto P_1 + P_2, [t_3; t_4) \mapsto P_2\} \,, \quad (7.4)$$

$$\{[t_1; t_3) \mapsto P_1\} \setminus \{[t_2; t_4) \mapsto P_2\}$$
$$= \{[t_1; t_2) \mapsto P_1, [t_2; t_3) \mapsto P_1 - P_2, [t_3; t_4) \mapsto -P_2\} \,. \quad (7.5)$$

Defining the 'subset' and 'subset-or-equal' relationships of individual mappings is equally feasible:

$$[t_1; t_2) \mapsto P_1 \subset [t_3; t_4) \mapsto P_2 \quad \Leftrightarrow \quad t_1 > t_3 \,\wedge\, t_2 < t_4 \,\wedge\, P_1 < P_2 \,, \quad (7.6)$$
$$[t_1; t_2) \mapsto P_1 \subseteq [t_3; t_4) \mapsto P_2 \quad \Leftrightarrow \quad t_1 \geq t_3 \,\wedge\, t_2 \leq t_4 \,\wedge\, P_1 \leq P_2 \,. \quad (7.7)$$

The Universal Agent software wraps one timespan-to-power mapping in a `Requirement` object, as depicted in Fig. 7.1. This class connects power messages to forecasts and contains factory methods to create an LPEP message object from a forecast and vice versa. `Requirement` objects also allow us to keep track of which requirement originated at the local node and which was sent by another agent, which is crucial for solving a disequilibrium.

7.2 The Combinatorial Demand-Supply Problem

In order to solve the (unbalanced) power balance, it makes use of a `PowerBalance-SolverStrategy`. Classes implementing this interface represent an actual algorithm for solving the power balance's disequilibrium in a given interval.

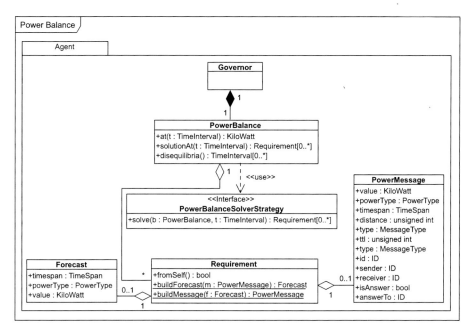

Figure 7.1: The Universal Smart Grid Agent's internal power balance

This disequilibrium, denoted by the power value P_0 in the time interval $\tilde{t}_0$, has been forecasted by an agent:

$$r_0 : [t_{0,1}; t_{0,2}) \mapsto P_0 \quad \Leftrightarrow \quad r_0 : \tilde{t}_0 \mapsto P_0 . \tag{7.8}$$

This agent formulates a request as an LPEP message, which, through selective broadcasting,[2] reaches other agents. These agents then check—and, in fact, must check—whether they can help to solve the disequilibrium, and if so, answer with individual LPEP messages:

$$r_1 : \tilde{t}_1 \mapsto P_1, r_2 : \tilde{t}_2 \mapsto P_2, \ldots, r_i : \tilde{t}_i \mapsto P_i , \tag{7.9}$$

$$\forall P_i, i \neq 0 : |P_i| \leq |P_0| , \tag{7.10}$$

$$\forall \tilde{t}_i, i \neq 0 : t_{i,1} \geq t_{0,1} \wedge t_{i,2} \leq t_{0,2} . \tag{7.11}$$

---

[2]Cf. Section 6.2.

The first and foremost task of the solver is now to find any combination of mappings, $\tilde{t}_1 \mapsto P_1, \ldots, \tilde{t}_i \mapsto P_i$, such that the initial disequilibrium P_0 is within the whole time interval $\tilde{t}_0$. We can therefore define the main goal of the solver as:

$$\sum_i b_i r_i \subseteq r_0 \ , \ i \neq 0, b_i \in \{0,1\} \ . \tag{7.12}$$

In reality, the nature of the power grid leaves a certain margin for over- or undersupply, which depends on the size of the grid. Remember from Section 4.1 that each agent features a *constraints module*, i.e., a software module that—among other things—supplies this margin. For the definition of the solver, it is denoted by P_C.

Furthermore, each LPEP message contains an accumulated distance value. This distance value is the sum of all impedances of all lines the message, and thus, the potential power transmission, travels. Since the LPEP models the power grid as an overlay network in the communications networks; hence, a communications-connection between two agents also corresponds to a power line. The higher the (accumulated) impedance, the higher the line losses of the power transmission. The second task of the solver is thusly: If more than one solution to a power disequilibrium exists, it must choose that with the lowest overall line loss. Let

$$\mathrm{d}(r_i) : r_i \mapsto \mathbb{R} \tag{7.13}$$

be a function the returns the accumulated distance of the requirement r_i. Using the function, we can define the secondary goal of the solver as:

$$\min \sum_i b_i \mathrm{d}(r_i), \ i \neq 0, b_i \in \{0,1\} \ . \tag{7.14}$$

The combinatorial problem the solver must find a solution to is therefore strongly reminiscent of the 0-1 knapsack problem (Dantzig et al., 2007). Since this is an optimization problem, numerous approaches present themselves, including utilizing the multipart evolutionary algorithm described in Section 5.2. However, neither the power values requested or offered by other agents, nor the associated time intervals, need to be subdivided to justify requiring the complete real numbers domain, $\mathbb{R}$, if another agent's request is subdividable at all. Therefore, modeling requests in terms of Boolean equations and solving the power disequilibrium in the Boolean domain provides an efficient approach to the local part of the demand-supply calculation.

7.3 A Boolean Model of Demand and Supply

Structure and Operation

The modelling of the local demand-supply calculation expresses the key problem in the calculation: A solution to the problem can be reduced to choosing the right combination of requirements—or, if possible, partial requirements—from other agents. We can therefore express the acceptance or rejection[3] of a requirement by modeling it in the Boolean domain.

However, we initially face a multi-valued problem: A requirement is comprised of any power value, P_i, as well as the corresponding time interval, $\tilde{t}_i = [t_{i,1}; t_{i,2})$. Therefore, at this point, we cannot use a single variable to represent it in the demand-supply calculation.

To break down the multi-valued problem into its discrete parts, we need to split each offer into its atoms.[4] The size of each atom is determined by the overall set of requirements the specific calculation is made of. This also includes the original request that is part of the power balance as well and which is denoted by the index 0, i.e., $[t_{0,1}; t_{0,2}) \mapsto P_0$. The size of the atoms is calculated from the vector of all power values,

$$\boldsymbol{P} = (|P_0|, |P_1|, \ldots, |P_i|, |P_C|) \;, \tag{7.15}$$

where P_C is the allowable power deviation as given by the constraints module,[5] i.e., the solution must match $P_0 \pm P_C$.

Further, all timespan sizes,

$$\boldsymbol{t} = (t_{0,2} - t_{0,1}, t_{1,2} - t_{1,1}, \ldots, t_{i,2} - t_{i,1}) \;, \tag{7.16}$$

naturally also influence the size of the atoms, which is their respective *Greatest Common Divisor* (GCD):

$$\Delta P = \gcd(\boldsymbol{P}) \;, \tag{7.17}$$
$$\Delta t = \gcd(\boldsymbol{t}) \;. \tag{7.18}$$

---

[3]Cf. the semantics of the LPEP for actions taken by the agent when the power balance is not solvable.

[4]The word 'atom,' from ancient Greek ἄτομον (átomon), means 'indivisible' and thus lends itself very well to the definition of the basic Boolean variables the solver uses.

[5]Cf. Fig. 4.1.

The GCD can be calculated requirement-by-requirement due to the associative law:[6]

$$\gcd(x, \gcd(y, z)) = \gcd(\gcd(x, y), z) \ . \tag{7.19}$$

The application of the GCD creates a raster of $n \cdot \Delta P \times m \cdot \Delta t$ atoms in which every requirement can be located. Thus, each agent's contribution to the power disequilibrium, i.e., its requirement, is deconstructed into a number of atoms.

Definition 7.1. *All requirements in the agent's power balance are deconstructed into a number of atoms. Each atom denotes a part of a requirement and references a time subinterval and a power subinterval in the power balance. The size of the time subinterval is Δt for all requirements; the size of each power subinterval is ΔP for all requirements. Each atom is described by a Boolean variable that expresses the origin of the requirement and the time and power interval in which it is located:*

$$x_{i, \tilde{t}, \tilde{P}} = \begin{cases} 1 & \textit{if the agent } i \textit{ influences the power grid in the time} \\ & \textit{subinterval } \tilde{t} \textit{ with power from the power subinterval} \\ & \tilde{P}, \\ 0 & \textit{otherwise.} \end{cases} \tag{7.20}$$

An example of this subdivision is shown in Fig. 7.2.[7] It contains five requirements—one forms the initial power disequilibrium, and four are (overlapping) responses from other agents—and illustrates how the $x_{i, \tilde{t}, \tilde{P}}$ variables reference the parts of the respective requirement.

Using these atoms, we can express the semantics of a `Requirement` object: A requirement expresses a power delta within a certain time interval. Specifically, it expresses a number of power deltas that are available for the solver to choose from in the time interval. Agents responding to a request can do so with a single power value, an interval of power values, or even offer multiple power intervals.[8]

---

[6]Theoretically, the GCD for a given set of requirements could be calculated continuously, as new requirements arrive.

[7]NB. that this particular power balance is unsolvable in its depicted state: Only the time subinterval 5 has sufficient offers to form a partial solution. It has intentionally been crafted this way to be clearly arranged while still being a potential state of an agent's power balance. After all, the solver algorithm can always fail due to a shortage of responses.

[8]Cf. Section 6.2.

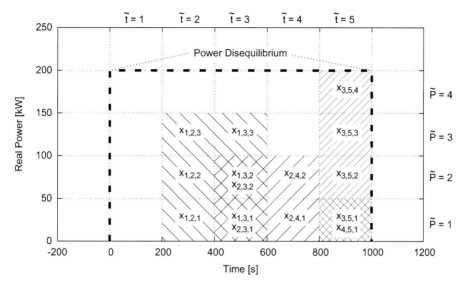

Figure 7.2: An example of a power balance state after the discretization

This information is obviously important to the solver. The *acceptance function* expresses this particular information of a requirement.

Definition 7.2. *A requirement can be a response to a request formulated via the LPEP. The response—i.e., the LPEP message—expresses the flow of power in a certain time interval. Depending on the intention of the agent the requirement originated on, the requirement can be accepted completely, in certain power quantities, or not at all. In the Boolean domain, this semantic is expressed through the* acceptance function *of each requirement:*

$$
\mathrm{r}_i(\boldsymbol{x}_{i,\tilde{t},\tilde{P}}) = \begin{cases} 1 & \textit{if } \boldsymbol{x}_{i,\tilde{t},\tilde{P}} \textit{ denotes a valid interval for accepting the} \\ & \textit{requirement from agent } i, \\ 0 & \textit{otherwise.} \end{cases} \tag{7.21}
$$

The atoms defined in Definition 7.1 are now used by the solver to create the characteristic function of each requirement's acceptance function. A requirement is not simply based on its atoms: They possess a certain coherence. An agent typically indicates, through its demand notification or offer notification,[9] what divisions into shares are possible. From this, the Boolean power balance solver creates conjunctions that specify the *acceptance function*. Each conjunction corresponds to a response and describes a possible acceptance of a requirement.

Section 6.2 described how several messages can make up one response: Through the characteristic function, the solver collapses these distinct messages back into one unified response by creating a conjunction for every possible valid interval as defined in Eq. (7.21). This is also required by the semantics of the protocol. Thus, for all time subintervals $\tilde{t}$ of size Δt within which the offer is valid, a conjunction for each allowable quantity of power, expressed as one or more power subintervals $\tilde{P}$ of size ΔP, is created. Thus, we can now generally express a requirement's acceptance function:

$$
\mathrm{r}_i(\boldsymbol{x_{i,\tilde{t},\tilde{P}}}) = \bigwedge_{\tilde{t} \in r_i, \tilde{P} \in r_i} \bar{x}_{i,\tilde{t},\tilde{P}}
$$

$$
\vee \left(\bigwedge_{\tilde{t} \in r_i, \tilde{P} \in r_i} x_{i,1,1} \wedge \bar{x}_{i,1,2} \wedge \bar{x}_{i,1,3} \wedge \cdots \wedge \bar{x}_{i,\tilde{t},\tilde{P}} \right)
$$

$$
\vee \left(\bigwedge_{\tilde{t} \in r_i, \tilde{P} \in r_i} x_{i,\tilde{t},1} \wedge x_{i,\tilde{t},2} \wedge \bar{x}_{i,\tilde{t},3} \wedge \cdots \wedge \bar{x}_{i,\tilde{t},\tilde{P}} \right)
$$

$$
\vee \cdots \vee
$$

$$
\bigwedge_{\tilde{t} \in r_i, \tilde{P} \in r_i} x_{i,\tilde{t},\tilde{P}} \; . \quad (7.22)
$$

The intermediate part of Eq. (7.22) depends on what choices the responding agent offers. However, the solver needs to create at least a number of conjunctions equal to $2 \cdot \frac{P_i}{\Delta P} \cdot \frac{t_{i,2} - t_{i,1}}{\Delta t}$ [10] since in the simplest case, a requirement is either completely accepted or not accepted at all:[11]

$$
\mathrm{r}_i(\boldsymbol{x_{i,\tilde{t},\tilde{P}}}) = \bigwedge_{\tilde{t} \in r_i, \tilde{P} \in r_i} x_{i,\tilde{t},\tilde{P}} \vee \bigwedge_{\tilde{t} \in r_i, \tilde{P} \in r_i} \bar{x}_{i,\tilde{t},\tilde{P}} \; . \quad (7.23)
$$

---

[9]Cf. Section 6.2.
[10]Cf. Fig. 7.2.
[11]This is simply the first and the last term in Eq. (7.22).

The BVs in $x_{i,\tilde{t},\tilde{P}}$ can be represented in a compact manner by a TVL. Remember Fig. 7.2 with its five responses; most[12] could actually offer more than two choices to the solver—accept completely, accept partially, or not accept at all—and such a possible concrete acceptance function is shown in Fig. 7.3.

Now that each non-initial requirement[13] is converted into a number of conjunctions of its atoms, the solver must model the actual request, i.e., the power disequilibrium. Since the order in which the requirements are accepted is not important, but the cover is, the solver models this using a symmetric function. A symmetric function's value at any n-tuple is the same at every permutation of that n-tuple; the function does not depend on the order of its variables, but only on the number of set or unset variables. A symmetric function $S^n(x_{i,\tilde{t},\tilde{P}})$ is equal to 1 iff exactly n of its variables are equal to 1, for every permutation of the assignment of its argument vector.

The solver creates m—where m denotes the amount of time subintervals that have been created through the application of the GCD[14]—symmetric functions, one for each time subinterval $\tilde{t}$ of size Δt. Thus, each symmetric function describes $\frac{1}{m}$ of the disequilibrium. The argument of the respective symmetric function are those parts of each requirement's characteristic function for the time subinterval the symmetric function is constructed for. This is indicated by the subscript k of the symmetric function; each symmetric function describes the k-th power subinterval. Stringently, the function's argument vector is written as $x_{i,\tilde{t}=k,\tilde{P}}$ to express that only the atoms of the k-th power subinterval are a part of the argument vector.

The number of set bits corresponds to the power disequilibrium: If the power disequilibrium amounts to $n \cdot \Delta P$ kW, with $n \leq |x_{i,\tilde{t}=k,\tilde{P}}|$, the symmetric function has n 1-bits and $|x_{i,\tilde{t}=k,\tilde{P}}| - n$ 0-bits. This implies that the responses the requesting agent received can actually solve the power disequilibrium. If $|x_{i,\tilde{t}=k,\tilde{P}}| < n$ is true for any k, no symmetric function can be created and the disequilibrium cannot be solved. The definition of the symmetric function is:

$$S_k^n(x_{i,\tilde{t}=k,\tilde{P}}) = \begin{cases} 1 & \text{if } n \text{ variables in } x_{i,\tilde{t}=k,\tilde{P}} \text{ are } 1, \\ 0 & \text{otherwise,} \end{cases} \qquad k = 1, 2, \ldots, m .$$

(7.24)

Relating to Fig. 7.2, the symmetric functions represent 'column-wise' slices of the respective requirements.

---

[12]With the exception of r_5

[13]I.e., all responses

[14]Cf. Eq. (7.17).

$$r_2(\boldsymbol{x}_{i,\tilde{t},\tilde{P}}) = \bar{x}_{2,3,1} \wedge \bar{x}_{2,3,2} \wedge \bar{x}_{2,4,1} \wedge \bar{x}_{2,4,2}$$
$$\vee\, x_{2,3,1} \wedge x_{2,3,2} \wedge \bar{x}_{2,4,1} \wedge \bar{x}_{2,4,2}$$
$$\vee\, x_{2,3,1} \wedge x_{2,3,2} \wedge x_{2,4,1} \wedge x_{2,4,2}$$

$x_{2,3,1}$	$x_{2,3,2}$	$x_{2,4,1}$	$x_{2,4,2}$	$r_4(\boldsymbol{x}_{i,\tilde{t},\tilde{P}})$
0	**0**	**0**	**0**	**1**
1	0	0	0	0
0	1	0	0	0
0	0	1	0	0
0	0	0	1	0
1	**1**	**0**	**0**	**1**
1	0	1	0	0
1	0	0	1	0
0	1	1	0	0
0	1	0	1	0
0	0	1	1	0
1	1	1	0	0
1	1	0	1	0
1	0	1	1	0
0	1	1	1	0
1	**1**	**1**	**1**	**1**

Figure 7.3: An example of the acceptance function

The solution set to the disequilibrium in $\tilde{t}$ consists of the exact covers of the symmetric function and the acceptance functions at $\tilde{t}$:[15]

$$C_k(\boldsymbol{x}_{i,\tilde{t}=k,\tilde{P}}) = S_k^n(\boldsymbol{x}_{i,\tilde{t}=k,\tilde{P}}) \wedge \bigwedge_{i \in I'} r_i(\boldsymbol{x}_{i,\tilde{t}=k,\tilde{P}}) \;. \qquad (7.25)$$

Here, the set I' is the set of all agents that have sent a response to the local agent's request. The complete solution set therefore consists of the complete cover, which is:

---

[15]NB. the subscript.

$$C(\boldsymbol{x}_{i,\tilde{t},\tilde{P}}) = \bigwedge_k C_k(\boldsymbol{x}_{i,\tilde{t}=k,\tilde{P}}) . \tag{7.26}$$

XBOOLE[16] can represent these equations in software and calculate the solution set efficiently. I.e., the Universal Agent uses XBOOLE as underlying software to implement the solver. The solver represents the acceptance functions as TVLs in *Orthogonal Disjunctive/Antivalent* (ODA) form, such as the example in Fig. 7.3. The symmetric functions are represented in the same manner and are even easier to generate: Their TVL consists of all permutations $n \times 1$ and $(|\boldsymbol{x}_{i,\tilde{t}=k,\tilde{P}}| - n) \times 0$ for the k-th symmetric function.[17]

XBOOLE's actual work consists of a number of intersections[18] of the TVLs. First, XBOOLE creates a TVL of all acceptance functions:

$$R = \bigcap_{i \in I', \tilde{t}, \tilde{P}} r_i(\boldsymbol{x}_{i,\tilde{t},\tilde{P}}) , \tag{7.27}$$

as well as one TVL representing all symmetric functions:

$$S = \bigcap_{k=1}^{m} S_k^n(\boldsymbol{x}_{i,\tilde{t}=k,\tilde{P}}) . \tag{7.28}$$

The intersection of two disjoint TVLs—i.e., two TVLs whose variable sets are disjoint—yields their Cartesian product,[19] i.e.,

$$\texttt{ISC}(P,Q) \quad \Leftrightarrow \quad P \times Q \quad \text{iff } \texttt{SV\_ISC}(\texttt{SV\_GET}(P), \texttt{SV\_GET}(Q)) = \emptyset . \tag{7.29}$$

The operator $\texttt{SV\_GET}(P)$ returns the set of variables of the TVL P.[20] The complete cover then constitutes the final intersection:

$$C = S \cap R . \tag{7.30}$$

---

[16]Cf. Section 2.5.

[17]NB. that this assumes a full cover is desired, i.e., that the original request should be matched completely. For practical purposes, this limits the solution space unnecessarily and therefore a number of 0 or 1 equal to $\frac{P_C}{\Delta P}$ are replaced by a '−' that allows under- or overmatching by $\pm P_C$. For exactly this reason, Eq. (7.15) includes P_C in the vector of power values.

[18]$\texttt{ISC(tvl1, tvl2)}$: Cf. Table 2.2.

[19]Cf. Bochmann and Steinbach (1991).

[20]Cf. Bochmann and Steinbach (1991, p. 14).

The TVL C that denotes the complete cover contains the complete solution set to the demand-supply calculation. If it contains more than one solution, i.e., $|C| > 1$, the agent must still choose among them. Ideally, if it is not restricted by contract or other constraints given by the agent's constraints module, it will try to minimize the line loss, and, therefore, prefer offers with a lower distance value.[21]

Let us assume that the function $\mathrm{d}(r_i)$ returns the distance value of a requirement r_i. This allows the ordering of all (accepted) requirements:

$$r_i \leq r_{i'} \quad \Leftrightarrow \quad \mathrm{d}(r_i) \leq \mathrm{d}(r_{i'}) \; . \tag{7.31}$$

C contains all variants that solve the power disequilibrium; from the TVL the solver can also extract the respective requirements that form the possible solutions. Through Eq. (7.31), we can calculate the sum of all distances of all requirements forming a solution. Thus, the TVs in C can be sorted by distance and the one with the lowest distance is returned. This is then the optimal solution to the power disequilibrium.

Complexity Analysis and Comparison

In order to return the result TVL, the XBOOLE solver creates a number of intermediate TVLs: One for each response, containing the individual acceptance function, one TVL that combines all acceptance functions, one for each symmetric function corresponding to a time subinterval, and finally one that combines all symmetric functions. However, since the TVLs for the individual acceptance functions as well as for the individual symmetric functions are only needed to create the Cartesian product, i.e.,

$$R_1 \times R_2 \times \cdots \times R_n \quad \Leftrightarrow \quad \mathrm{ISC}(R_1, \mathrm{ISC}(R_2, \mathrm{ISC}(\ldots, R_n))) \; , \tag{7.32}$$

only at most 4 TVLs are used by the solver at a given time.

The size of the individual TVLs that represent the symmetric functions is a combinatorial classic, specifically, how often one can choose k elements from a set of the size n, disregarding the order. Here, the size of the set is the size of the argument vector of the symmetric function; the number of elements that should be chosen is n, i.e., the number of set bits in the argument vector. Therefore, we can express the size of the TVL through a binomial coefficient.

---

[21]Cf. Definition 6.3.

Table 7.1: Ternary Vector List representing S_5^4 of the example power balance

$x_{3,5,1}$	$x_{3,5,2}$	$x_{3,5,3}$	$x_{3,5,4}$	$x_{4,5,1}$
0	1	1	1	1
1	0	1	1	1
1	1	0	1	1
1	1	1	0	1
1	1	1	1	0

Thus, each of the m individual symmetric functions that correspond to one time subinterval contains a number of TVs equal to:

$$|S_k| = \mathtt{NTV}(S_k) = \binom{|\boldsymbol{x}_{i,\tilde{t}=k,\tilde{\boldsymbol{P}}}|}{n} = \frac{|\boldsymbol{x}_{i,\tilde{t}=k,\tilde{\boldsymbol{P}}}|!}{n! \cdot (|\boldsymbol{x}_{i,\tilde{t}=k,\tilde{\boldsymbol{P}}}| - n)!} \ . \quad (7.33)$$

We can verify this formula with the example shown in Table 7.1 that shows the TVL representing $S_5^4(\boldsymbol{x}_{i,5,\tilde{P}})$ the solver would create for the example power balance depicted in Fig. 7.2. Here, we can calculate $|S_5^4| = \binom{5}{4} = 5$.

The size of each TVL representing the individual offer's acceptance function depends on the number of power values offered, but contains at least one row for 'accept completely' and one for 'accept not at all':

$$|R_i| \geq 2 \ . \quad (7.34)$$

In comparison to other structures representing binary functions, such as BDDs, the TVL approach of XBOOLE proves to be more efficient in terms of space complexity. Remember from Section 2.5, that the space complexity of a BDD depends on the ordering of variables. The set of variables naturally also influences the size of any TVL. We can therefore use the $\mathtt{SV\_SIZE}(P)$ operation to initially determine the number of variables:

$$nv = \mathtt{SV\_SIZE}(R) \ . \quad (7.35)$$

Ideally, with good variable ordering, a BDD with $nv = 2k + 2$ variables can be compressed to be minimal.[22] Then, the BDD will store

---

[22]Cf. Bryant (1986).

$$|\boldsymbol{v}| = 2k + 2 \tag{7.36}$$

vertices. Under ideal circumstances, a TVL can represent an entire function with only one TV,[23] requiring a number of ternary elements equal to:

$$nte = nv \ . \tag{7.37}$$

This might seem like an academic consideration, however, we will later see that through optimization, we can indeed arrive at this space requirement formula for many functions the solver considers.

Symmetric functions force us to evaluate the worst-case space complexity since they resist a data structure's approach to a compact representation. In general, using k variables, we can describe 2^{2^k} functions. Each vertex of a BDD is also the root of a subfunction; a BDD therefore describes a set of functions. In order to construct the worst case, a number of variables equal to $nv = k + 2^k$ must be given. Otherwise, the BDD can be compressed. Thus, the worst-case number of vertices of a BDD with $nv = k + 2^k$ variables is:

$$|\boldsymbol{v}| = 2 \cdot 2^{2^k} - 1 \ . \tag{7.38}$$

The maximum number of ternary values (i.e., elements) in a TVL[24] with nv variables then corresponds to:

$$nte = nv \cdot 2^{nv-1} \ . \tag{7.39}$$

To compare the space complexity of BDDs and TVLs, we must consider their actual memory requirements. A BDD is made of data type definitions in the form of this C struct:[25]

```
struct tree_node_t
{
    struct tree_node_t* high;
    struct tree_node_t* low;
    unsigned index;
    enum value_t value;     /* 0, 1, or X */
```

---

[23]Even more, the empty TVL represents the zero function.

[24]The TVL is then, in the worst case, a list of BVs.

[25]This excludes, as with the TVL, the size of the variable name. The translation to the C programing language is based on the record definition by Bryant (1986). It ignores padding issues deliberately.

```
    unsigned id;
    unsigned mark:1;
};
```

Thus, each vertex takes up 2 words to represent the edges of the tree, additional 2 words to store the `index` and `id` attributes, and, finally, 3 bits for `value` and `mark`. If we ignore padding, we can express the storage requirement of a BDD as:

$$D_{BDD} = |\boldsymbol{v}| \cdot (4D_W + 3) \text{ bits.} \tag{7.40}$$

On a recent x86_64 architecture, this would mean 64 bits per word. In contrast, a ternary value uses only 2 bits,[26] regardless of the processor architecture, sizing the TVL at:

$$D_{TVL} = 2nte \text{ bits.} \tag{7.41}$$

The smallest BDD that consists of one root and two terminal nodes will require:

$$\min(D_{BDD}) = 3 \cdot (4D_W + 3) = 777 \text{ bits}, \tag{7.42}$$

enough space for 388 ternary elements.[27] Considering the worst case, the BDD becomes the better choice only for $nv \geq 521$ variables; at this point, $D \approx 8 \times 10^{146}$ GB will be required—too much by all means.[28] Again, we will later see that we can avoid the worst case altogether, taking full advantage of the compact storage structure of a TVL and the superior computational complexity of operations on TVLs, as we will assert below.

The value of the GCD obviously directly influences the number of variables and thus the size of the individual TVLs, as the occurrences of the variables n and m indicate. One might therefore assume that a model of the given problem that works on integers would greatly reduce the memory footprint and complexity of the solver. Such a structure is an EVMDD that can represent any p-valued function.[29] Let us define an offer as a p-valued function of two variables, $\tilde{t}$ and $\tilde{P}$. The function arguments denote the time and power interval for the

---

[26]Cf. Bochmann and Steinbach (1991, Chapter 6.1).

[27]NB. that this difference weights heavy in the best case in favor of the TVL, and even more if the space compression ability of the ternary values can be used.

[28]Replacing the GCD with an interval partitioning algorithm can lower the number of variables even further and avoids this critical margin.

[29]Cf. Section 2.5.

requirement from agent i. The function remodels the acceptance equation of a requirement, described in Eq. (7.21) as a pure Boolean function, now as a p-valued function:

$$r_i(\tilde{t}, \tilde{P}) = \begin{cases} \tilde{P} \cdot \Delta P & \text{if } t_{1,i} \leq \tilde{t}\Delta t \leq t_{2,i} \ , \\ 0 & \text{otherwise.} \end{cases} \qquad (7.43)$$

According to Nagayama and Sasao (2007), the memory size of an EVMDD is:

$$D_{EVMDD} = D_W \sum_{k=1}^{u} 2p_k \cdot w_k \text{ bits}, \qquad (7.44)$$

where u denotes the number of multi-valued variables and w_k the number of non-terminal nodes for a variable. Comparing the worst-case memory size of a TVL in Eq. (7.41) with that of a EVMDD in Eq. (7.44), one can easily conclude that the EVMDD is the more efficient data structure once the rasterization[30] leads to a certain number of variables for the acceptance functions and, correspondingly, the symmetric function.

However, the memory of an embedded appliance running the Universal Agent software should be well suited to store a number of kilobytes to accommodate the data structures. More interestingly, the runtime behavior of the XBOOLE-based solver is more favorable. We can assume that the solution to a demand-supply calculation is achieved by a combination of two EVMDD subtrees—one for the acceptance function and another one for the symmetric functions—, followed by a *satisfy-one* operation.[31] If we consider the runtime complexity of BDDs and, therefore, EVMDDs, listed by Bryant (1986), we conclude that the computational complexity of an EVMDD approach would be:

$$\mathcal{O}\left(|R|^2 \cdot |S|\right) + \mathcal{O}\left(|R| + |S|\right) \ , \qquad (7.45)$$

whereas the computational complexity of the intersection operation—$\text{ISC}(P, Q)$—of XBOOLE is:[32]

$$\mathcal{O}\left(|R| \cdot |S|\right) \ . \qquad (7.46)$$

---

[30]Cf. Eq. (7.17).
[31]Cf. Bryant (1986, Tab. 1).
[32]Cf. Bochmann and Steinbach (1991, p. 259).

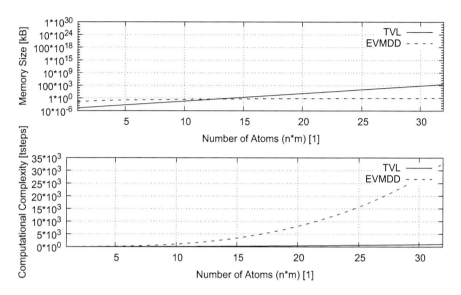

Figure 7.4: Memory size and computational complexity of XBOOLE-based and Edge-Valued Multi-valued Decision Diagram-based demand-supply solvers

Fig. 7.4 depicts the influence of n and m with regards to the TVL and EVMDD memory size and computational complexity. In comparison, while the EVMDD approach is the more space efficient data structure, the TVL-based modelling approach is by far the faster one.

This comparison serves as a transparent side-by-side analysis of a TVL-based versus an EVMDD-based approach to the solver. However, if an implementing algorithm followed Eqs. (7.24) to (7.28) and (7.30) by the letter, the solution would not be the optimal one. Every intersection except for the last one creates the Cartesian product of the two TVLs, since they are disjoint.[33] No two TVLs representing their respective acceptance functions share a variable, since an offer is unambiguously identified by the variable set that makes up its atoms per Eqs. (7.17) and (7.20). The symmetric functions that serve to model the initiating request add another burden: They are, by their nature, immune to simplification using the ternary value. I.e., at no time will any TVL representing a symmetric function contain a '$-$.'

---

[33]Cf. Eq. (7.29).

However, optimization is possible by exploiting two properties of the model. First, the intersection is commutative, and so is XBOOLE's $\texttt{ISC}(P, Q)$ operator. Instead of creating the complete acceptance function, R, directly for the intersection with the set of all symmetric functions, as Eqs. (7.27) and (7.30) suggest, we can compute the intersection iteratively for each acceptance function:

$$R_i \cap (\ldots \cap (R_2 \cap (R_1 \cap S))) \ . \tag{7.47}$$

This alleviates the solver from the task of creating the complete set R, but yields the same result as the naïve operation $R \cap S$.

The second property of the solver's model helps us to reduce the size of the set for which all symmetric functions, S, are equal to 1. It is also not necessary to create S completely. Since each row in a TVL representing a particular $S_{\tilde{t}}$ is a permutation of the first one, we can deduce the next TV in $S_{\tilde{t}}$ from the current one. Table 7.2 illustrates that one can easily establish an order within the function BVL for any symmetric function. We can thus define and easily implement[34] the following operations for a symmetric function's function BVL:

FIRST($\boldsymbol{sbv}$) Creates the first BV in the BVL $\boldsymbol{sbv}$ is a part of.

NEXTPERMUTATION($\boldsymbol{sbv}$) Derives the next permutation from $\boldsymbol{sbv}$, i.e., it creates the next BV of the BVL for the symmetric function $\boldsymbol{sbv}$ is also a part of.

LAST($\boldsymbol{sbv}$) Creates the last BV in the BVL $\boldsymbol{sbv}$ is a part of.

We can now permute a BV and, thus, do not need to store the complete TVL for a symmetric function. This allows us to compress the function TVLs for the symmetric functions for each time-subinterval, $S_{\tilde{t}}$, to the size of one—the current—BV.[35] I.e., instead of working with the whole TVL for each time-subinterval's symmetric function, $S_{\tilde{t}}$, we now only require the current BV of the TVL. The current BV for the symmetric function for all time-subintervals, S, is then simply a concatenation of the respective BVs.[36] The permutation

---

[34]These functions are indeed part of a number of standard libraries, such as C++'s *Standard Template Library* (STL) (cf., e.g., cppreference.com Contributors (2015)). The implementation in the STL works by exploiting the lexicographic ordering of the elements of a BV, i.e., by defining $0 \prec 1$ (*false* precedes *true*).

[35]Remember Eq. (7.37): Now, we can indeed represent all symmetric functions with only one binary vector, making the TVL-based solver also the most space-efficient solution.

[36]This concatenation is created by the intersection function, $\texttt{ISC}(P, Q)$.

Table 7.2: Binary Vector List for an example symmetric function, $S^3(\boldsymbol{x})$

x_1	x_2	x_3	x_4	x_5
0	0	1	1	1
0	1	0	1	1
0	1	1	0	1
0	1	1	1	0
1	0	0	1	1
1	0	1	0	1
1	0	1	1	0
1	1	0	0	1
1	1	0	1	0
1	1	1	0	0

must still obey the boundaries that are introduced through the individual time-subintervals, i.e., it must generate a permutation for each current BV in the respective $S_{\tilde{t}}$ and not over the whole S. Therefore, we must construct a function that, given the current TVL of each time-subinterval's symmetric function, i.e., a vector of TVLs, creates the next valid permutation and returns the next valid BV in S. This method is called NEXTSYMMETRICFBV($\boldsymbol{S_{\tilde{t}}}$). Notice its argument type: A vector of TVLs. Algorithm 7 outlines the function's modus operandi.

We can thus construct a function that returns the next permutation of the TVs in S without having to create S completely. Therefore, we can note the solver's final form in Algorithm 8.

This final version starts with the calculation of the GCD and the construction of all relevant TVLs: First, R contains a TVL for each response—i.e., it is a collection of TVLs—, then $\boldsymbol{S_{\tilde{t}}}$ is introduced to hold a TVL for each time subinterval. However, each respective TVL, $S_{\tilde{t}}$, only contains one BV, which is the first valid permutation of the respective symmetric function's TV.

Its main part is a loop in which all permutations in S are generated, using Algorithm 7. The resulting TVL S of each permutation is then subsequently intersected with each response TVL, R_i,[37] and the result appended[38] to the final solution TVL. The loop ends when all valid permutations have been generated.

---

[37]Cf. Eq. (7.47).
[38]Using the concatenation operator, CON(P, Q)

Algorithm 7 Calculation of the next valid permutation for the Binary Vector of the symmetric function

---

procedure NEXTSYMMETRICFBV($\boldsymbol{S_{\tilde{t}}}$)

    for $k = |\boldsymbol{S_{\tilde{t}}}|, 1$ **do**   ▷ Iterates over the function TVLs.

        $S_{\tilde{t}} = \boldsymbol{S_{\tilde{t},k}}$

        $\boldsymbol{sbv} \leftarrow S_{\tilde{t},1}$ ▷ First BV in the respective TVL

        if $\boldsymbol{sbv} \neq$ LAST($\boldsymbol{sbv}$) **then**

            $\boldsymbol{sbv} \leftarrow$ NEXTPERMUTATION($\boldsymbol{sbv}$)

            break

        else if $k = 1$ **then**

            return $\emptyset$ ▷ No more permutations possible

        else

            $\boldsymbol{sbv} \leftarrow$ FIRST($\boldsymbol{sbv}$)

        end if

    end for

    $S \leftarrow S_{\tilde{t},1}$

    for $k = 2, |\boldsymbol{S_{\tilde{t}}}|$ **do**

        $S \leftarrow S \cap S_{\tilde{t},k}$

    end for

    return S

end procedure

The best solution in the resulting TVL—if it is not empty, in which case no solution exists—is determined by sorting its TV. The basis for the sorting operation is Eq. (7.31), from which we can deduce that each TV in the solution TVL has an accumulated distance value.[39] Thus, the first TV[40] indicates the best solution. The solver now needs to map the TV back to the respective responses and to return the set of solution responses.

Fig. 7.5 compares the data volume required by the Universal Agent against the line loss that is avoided by its operation. The plot assumes that the node on the last hop answers and that the GCDs of the Δt and ΔP atoms is 1, which means the worst-case size of the TVLs' set of variables. Through the optimized version of the solver in Algorithm 8, it is possible to keep the total amount of data required to arrive at a solution—including both, the volume of all messages transmitted and the size of the TVLs at the requesting node—below

---

[39]I.e., the LPEP distance metric of the underlying message; cf. Algorithm 6.

[40]Which is retrieved using STV(*solution*, 1)

Algorithm 8 The Universal Smart Grid Agent's central solver procedure

---

procedure SOLVE(*request, responses*)
    $(\Delta P, \Delta t) \leftarrow \text{GCD}(request, responses)$
    $R \leftarrow \text{CREATERESPONSETVLs}(responses, \Delta P, \Delta t)$
    $\boldsymbol{S_{\tilde{t}}} \leftarrow \text{CREATEREQUESTTVLs}(request, responses)$
    $S \leftarrow \emptyset$
    $C \leftarrow \emptyset$
    repeat
        $S = \text{NEXTSYMMETRICFBV}(\boldsymbol{S_{\tilde{t}}})$
        $subSolution \leftarrow S$
        for all $R_i \in R$ **do**
            $subSolution \leftarrow subSolution \cap R_i$
        end for
        if $C = \emptyset$ **then**
            $C \leftarrow subSolution$
        else
            $C \leftarrow \text{CON}(C, subSolution)$
        end if
    until $S = \emptyset$
    if $|C| > 0$ **then**
        $C \leftarrow \text{SORT}(C)$ ▷ Per Eq. (7.31)
        return SELECTRESPONSES(C) ▷ Returns responses from solution
vector.
    else
        return $\emptyset$
    end if
end procedure

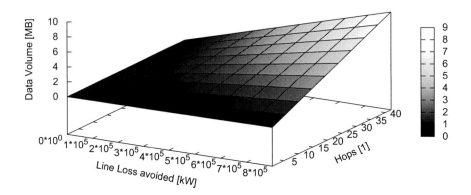

Figure 7.5: Data volume used compared to line loss avoided

10 MB while scaling up to quantities of 800 MW, which would include the whole capacity of a classic power plant such as one that uses bituminous coal, or even a nuclear power plant.[41]

Fig. 7.5 is the foundation from which we can arrive at a general metric to evaluate the performance of the Universal Agent, as well as other approaches.

7.4   Evaluation of Efficiency

Any proposal must show, at least in theory, that it improves the present situation. The Universal Grid Agent makes no difference in this regard. In order to quantify the impact of the agent approach, we need to define a metric. Since one of the pillars of the solution proposed in this thesis is to model the power grid in communication networks, comparing bytes and power transmitted suggests itself. Remember the metrics proposed by Bush (2014) that were introduced in Section 2.1. We define the *effect* of a distributed demand-supply calculation in terms of bytes it requires to transport a certain amount of power as the *data effect*:[42]

---

[41]Cf. Table 3.1.

[42]The symbol κ is the first letter of the ancient Greek word κράτος, which means 'effect,' 'potency,' or 'might.'

$$\kappa = \frac{W}{D} \left[\frac{\text{kWh}}{\text{kB}}\right] . \tag{7.48}$$

This metric is, in fact, applicable for every solution and not specific to the Universal Agent. In this case, specifically, we can evaluate the minimum value for κ by assuming the following parameters:

1. The lowest quantity transmittable is $1\,\text{kW}$. This stems immediately from the design of the LPEP.

2. The smallest time interval usually encountered is $10\,\text{min}$ long. This, too, follows from the proposition made by the LPEP semantics.

The four-way handshake of the LPEP causes at least $1120\,\text{B} \approx 1.1\,\text{kB}$ to travel, when assuming JSON notation. The smallest effect of one link is therefore:

$$\kappa_{1l} = \frac{1 \cdot \frac{1}{6}}{1.1} \approx 0.15\,\text{kWh/kB} . \tag{7.49}$$

Fig. 7.6 compares the active power output of the 'Bare Hill Wind Farm' with the backfeed of the 'White Hill Springs Substation.' For a more realistic application of the data effect, we assume that the pursued goal is to reduce the substation's backfeeding and the wind farm therefore is required to offer its active power production or be curtailed.

In the first interval, the wind farm offers $P_1 = 10\,739\,\text{kW}$. Consulting Table 6.2, we know that the offer travels on up to 93 links; from Fig. 3.2 we can conclude that there will be at most 27 answers traveling via 169 hops in total, giving:

$$\kappa_1 = \frac{10739}{6 \cdot (93 \cdot 0.268 + 169 \cdot 0.804)} \approx 11.13\,\text{kWh/kB} , \tag{7.50}$$

iff all answers also constitute the solution set.[43] The data effect metric can, together with a reference situation as described in Section 3.2, be used to compare protocols with each other. The more Watts per Bit are transmitted, the more a effective a protocol is.

However, this does not describe the efficiency of an approach. Managing volatile power generation and consumption does not only mean to act on or

---

[43]Cf. Eq. (7.30).

react to this volatility as, e.g., curtailment is always possible, but also entails an increase in efficiency due to local generation and consumption: Power that does not need to be transmitted via longer distances suffers lower line losses than power that travels many kilometers of cable. Of course, this assumes that local generation as a reaction to increased demand or higher consumption due to temporally higher power production are possible.

We can therefore define the *data efficiency* in terms of line loss avoided per bytes transmitted:

$$\xi = \frac{\Delta P}{D} \left[\frac{\text{kW}}{\text{kB}} \right] \ . \tag{7.51}$$

Consulting the reference grid in Fig. 3.2 and assuming an average line length of 10 km for all overhead wires, we can calculate the line loss to 8470 kW. According to Eq. (7.51), this sets the data efficiency to:

$$\xi_1 = \frac{8470}{6 \cdot (93 \cdot 0.268 + 169 \cdot 0.804)} \approx 8.78\,\text{kW/kB} \ , \tag{7.52}$$

meaning that the operation of the Universal Agent was able to reduce the (active) line loss by 8.78 kW per kilobyte transmitted. Plotting this over the course of a day, we arrive at the second graph depicted in Fig. 7.6 that shows the data efficiency achieved by the operation of the Universal Agent.

The two metrics introduced in Eqs. (7.48) and (7.51) can serve to compare protocols, the quality of solvers, and approaches in general. They allow us to compare approaches that solve the problem from different angles, since all these approaches in the context of the smart grid naturally require a computer, and hence a certain volume of data.

Let us compare the approach outlined in this thesis, the Universal Smart Grid Agent, to that proposed by Inoue et al. (2014), which uses BDDs. These two are solutions to the same problem, but different in their characteristics: The Universal Agent is distributed and de-centralized, whereas the BDD-using solution requires only one node and can eschew network communication.

Both use the same grid, namely, an electric grid developed by Fukui University and *Tokyo Electric Power Company* (TEPCO). Inoue et al. (2014) briefly describe the network whose raw data is available through Fujimoto (2006). It models a typical Japanese distribution network. The grid is fed from 72 feeders and contains 468 switches. Including electrical and topological constraints, it allows for 1.5×10^{70} feasible configurations. Given pre-defined loads, the goal consists in finding a configuration with minimal line loss. Additionally, this thesis offers the data efficiency metric, which we will apply, too.

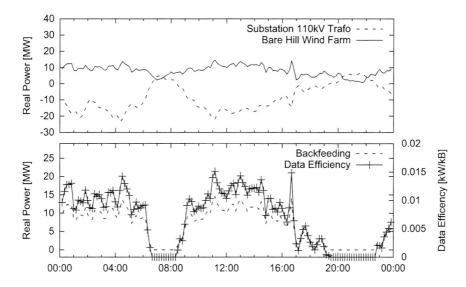

Figure 7.6: 'White Hill Springs Substation' backfeeding, 'Bare Hill Wind Farm' active power production, and data efficiency of grid-local consumption

Inoue et al. (2014) examine the Fukui-TEPCO grid at a low-load situation at 2 A.M., as well as at a high-load situation at 4 P.M. During high load, the BDDs that model the grid, as well as the constraints, require 100 MB of data volume. Modelling the grid in the network simulator, we can observe how the Universal Agent solves the same load situation. Here, each node and each section as well is represented by an agent. The sections with active and reactive power loads broadcast their requests until they reach the feeder nodes, which, in turn, reply with their offers. Those sections that represent switches choose the state of the switch according to whether they have relayed an acceptance acknowledgement notification[44] or not: If such a message was forwarded, the switch must be closed; otherwise, it will be opened. Since the LPEP features direct routing for all responses, no power will flow via these sections if no acceptance acknowledgement notification was relayed by these agents.

The results listed in Table 7.3[45] show that, for the high-load case, the

---

[44]Cf. Section 6.2.

[45]The line loss avoided is calculated against the theoretical line loss in the grid, which is, in turn, calculated according to the formulae presented by Nara et al. (1992).

Table 7.3: Comparison of the Universal Smart Grid Agent and Binary Decision Diagram approach on the Fukui-TEPCO power grid during high load

	BDD	Universal Agent
Line loss avoided (ΔP)	17 208 kW	17 208 kW
Compute Time	> 16 min	1 min 34 s (run time); < 11 min (simulated time)
Data Volume	100 MB	28.9 MB
Data Efficiency	0.168 kW/kB	0.581 kW/kB

decentralized agent concept requires less data volume than BDD approach and also takes less time to compute. The simulation run in which all simulated agents share a single thread finishes in less than 2 min, while the simulated time within the environment amounts to a little over 10 min.[46] This ten-minute-interval stems directly from the design of the LPEP.[47] In the high-load case, the Universal Agent therefore features a higher data efficiency than the approach by Inoue et al. (2014) based on BDDs.

However, the decentralized concept comes with a cost, which is the base data volume required to initially set up each agent instance. Fig. 7.7 shows the total data volume[48] required by all Universal Agents. The graph shows not only the base data volume required by the Universal Agent approach, but also indicates that the LPEP contributes the biggest share to the overall sum. We can also note that the XBOOLE-based power balance solver requires the least amount of data, although an instance runs on each requesting agent. This stems from the space efficiency of Algorithm 8, but is also due to the 72 feeders being able to offer exactly the amount of power required; thus, the size of the individual TVLs remains low.

The nature of the Universal Agent makes it unfitting for the low-load situation. Here, the BDD approach requires only 100 kB of data to run, whereas the Universal Agent's base data volume cannot be lowered: Each node and each section, regardless of their current load situation, is always represented by an

---

[46]Including network latency modelled as $X \sim \mathcal{U}[20\,\text{ms}; 250\,\text{ms}]$

[47]Cf. Section 6.2.

[48]This excludes the data base required by the Forecaster module of the agents since no historic data is available for the Fukui-TEPCO network.

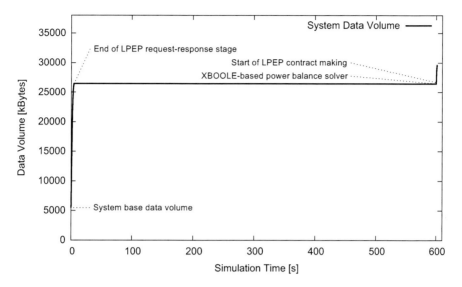

Figure 7.7: Total data volume by time required by the Universal Smart Grid Agent in the Fukui-TEPCO power grid during high load

instance of the Universal Agent.

The comparison thus served to show two results. It demonstrated that the metrics defined in Eqs. (7.48) and (7.51) can be applied to different approaches that try to improve efficiency in the smart grid. It also emphasized that the distributed, de-centralized concept that is the nature of the Universal Agent is suited best for complex situations with many actors, possibly of vastly different types, but falls short in simpler environments. For the future, we anticipate the smart grid to become an ever more complex environment due to distributed generation and volatile load and generation patterns. The operation of such a grid can be governed by the Universal Smart Grid Agent.

8   Conclusion

'Somewhere, there's always wind blowing or sun shining.' This optimistic maxim assumes that renewable energy sources, such as wind and photovoltaic, though volatile in nature, can become the main source of electrical power in our civilization. This thesis followed this assumption and discussed it from the point of availability, processing, and distribution of information. It reformulated the maxim into a question: 'Assuming enough primary energy is available, would the electric grid's power balance be a problem of information processing and distribution?'

The Universal Smart Grid Agent combines several technologies and builds upon them, thereby extending them. Its keystone is the notion of a distributed demand-supply calculation: Since the future power grid will be shaped by a vast amount of distributed generation along with power consumption that is distributed by nature, the necessary equilibrium between generation and consumption should also be the result of a distributed process. Each node in the power grid therefore actively contributes to this process, not just in terms of influencing the actual flow of real or reactive power, but also in terms of calculation.

This concept rests on three equally important pillars. The first is localized forecasting. Each node creates a forecast of its future power balance based on historical data drawn from its environment. It trusts the underlying idea that neither volatile power generation nor power consumption are chaotic processes, but that patterns inhere the behavior exhibited by photovoltaic arrays, wind farms, and consumers of every shade.

The forecaster module of the Universal Smart Grid Agent uses *Artificial Neural Networks* (ANN) to memorize existing, node-local patterns derived from historical data. It builds upon ANN patterns whose structure inherently recognizes time series and uses the network topology introduced by Jeff Elman to implement the forecaster. This thesis proposes and uses the data pipeline construct to lower the technical debt the application of machine learning impends to bring with it and separates data retrieval, logging, processing, forecaster configuration, and the actual forecasting, in order to make the Agent's architecture agnostic of its final deployment. The Agent employs the Multipart Evolutionary Training Strategy devised by Martin Ruppert, which has only published briefly before; this thesis adds a discussion and analysis of its features to the description of this remarkable algorithm.

This thesis finds itself in a contradictory position. A plethora of data is theoretically available and the advent of the *smart grid* heralds numerous sensors to be deployed, being accessible now or in the near future. But the future power grid mandates the additions of more data sources still: Patterns of date, time, wind speed, and wind direction can be contradictory, but other indicators of the general weather situation, such as barometric pressure, are usually missing since they were previously unneeded for the seamless operation of a wind farm. Node-local forecasting yields usable results of sometimes surprising accuracy and more advanced technologies, such as *Long Short-Term Memory* (LSTM) or deep learning techniques, will enhance the usefulness of it, given the node is sufficiently staffed with the necessary hardware. Here, specialized chipsets such as *Field-Programmable Gate Arrays* (FPGA) or *Application-Specific Integrated Circuits* (ASIC) can help to make this economically feasible. Such forecasting will be a valuable addition to meteorological weather forecasting, enhancing the confidence a grid operator—with the Agent being considerable as an automatic operator—can put into an estimation of the grid's state.

Data and information present at a node contribute towards the answer of the question initially posed, but are not the sole solution. Each node needs to act on its own information base, proactively working towards a power equilibrium in the grid. The active rather than reactive nature of this thesis' proposition gains it the designation of an agent in the stricter sense of artificial intelligence. Essential to an agent is its social component: A protocol for the nodes in the power grids—the agents—to converse constitutes the second pillar of the Universal Smart Grid Agent. Such a protocol must follow the Agent's active nature of behavior; its goal cannot be to provide a means to transport as many different pieces of information as possible or provide a networked query interface for sensors. Instead, this thesis proposed the *Lightweight Power Exchange Protocol* (LPEP) that is a protocol primarily in the sense of a set of behavioral rules and only secondary in the sense of a way to encode and transport data. The LPEP defines the social component the agent approach requires—in theory, any implementation can contribute to the distributed demand-supply calculation, not just the one proposed in this thesis.

The LPEP builds upon the well-known ISO/OSI stack model in order to accommodate different types of hardware and transmission technology required by the physical circumstances encountered in the field, be it long-wave, 3G, or WiFi radio connectivity or any sort of wired technology. Since its actual specification is motivated by the aforementioned set of behavioral rules, its intentions can be integrated into other, fully specified protocols, such as the *Open Smart Grid Protocol* (OSGP) or used with the *Common Information Model* (CIM). The LPEP's salient feature is that it re-models the power grid in the communications network as an overlay network. It assumes no a priori knowledge of the grid's nodes' purpose, but specifies mechanisms to ensure reliable and as efficient as possible communication between all agents.

It is safe to assume that the LPEP by itself will not see real-world deployment, but the rules that motivated its creation will.

After forecasting and communication, the third and final pillar of the Universal Smart Grid Agent concept is its heart piece: the actual solver for the demand-supply calculation that selects other nodes' offers and ensures the power equilibrium at each part of the electrical grid. Perhaps surprisingly, this forms the most dense and efficient part—memory- as well as computation-wise—of the approach outlined in this thesis. Its foundation is the excellent *Ternary Vector List* (TVL) arithmetic by Dieter Bochmann, Christian Posthoff, and Bernd Steinbach, that not only allows elegant modelling of demand and offers, but also proves to be the most efficient compared to other structural approaches in the Boolean domain, such as *Binary Decision Diagrams* (BDD).

With the Agent's heart piece in place, this thesis discussed the benefits of the agent approach. It introduced two metrics inspired by Steve Bush: Data effect and data efficiency. Data in the sense of bytes transmitted and processed by the individual agents measured against the change they induced in the power grid can serve as a metric to compare different approaches. They serve as a tool to answer the question posed initially, namely, whether the reliable and efficient integration of renewable energy sources is a problem of information distribution and processing and whether any software-based approach increases the efficiency of grid operation and provides energy gains by avoiding more line losses than the software requires power to operate.

In theory, the power grid's equilibrium can be viewed as a question of information distribution and processing and the operation as a software such as the Universal Smart Grid Agent can help to use power generated by volatile, renewable energy sources more, more efficiently, and more reliably.

Every approach must be tested, not only in restricted simulation, but also in larger-scale tests. This thesis also offered and discussed a simulation software approach whose principles can contribute to these efforts. Ultimately, only longer exposure of any technique to real-life circumstances can help to verify the feasibility of a proposed solution. While this world of the smart grid is still relatively new and exciting, it sometimes seems to clash with the established and durable world of traditional electrical engineering. From this point of view, any proposition from the domain of computer science has to be gauged against the required longevity and robustness of the power grid. A transformer's life time is roughly 30 years to 45 years. It remains a challenge to construct a smart grid whose software and hardware components can live up to an equal range.

Every conclusion to a thesis can, ultimately, only be the starting point for the next step of research. We must, therefore, as Alfred, Lord Tennyson formulated in his poem *Ulysses*, be

> "... strong in will
> To strive, to seek, to find, and not to yield."

A   Theses

1. Power generated from renewable energy sources forms an ever increasing share of the global energy mix.

2. For many regions, the go-to primary renewable energy sources are wind and solar radiation.

3. Wind and solar radiation are uncontrollable, volatile, primary energy sources since they depend on the weather.

4. The increasing percentage of wind farms and *Photovoltaic* (PV) power plants in the power grid increases the volatility of power generation.

5. Wind farms and PV installations feed into various, mostly lower, voltage levels of the power grid, thereby voiding the traditional, hierarchical architecture of the power grid.

6. Power generation is becoming increasingly decentralized while it traditionally was centralized and centrally-controlled.

7. Traditional, steam-based power plants are most efficient as base-load power plants that continuously feed in at their rated power output.

8. Power storage is, due to high costs, low efficiency, or space requirements, not available in quantities great enough to support a power generation based completely on volatile renewable energy sources.

9. The increasing share of volatile power generation necessitates flexible, load-following power plants and/or flexible power consumption to compensate this volatility.

10. Most traditional power plants cannot change their output at a high rate and cannot feed power below a certain minimum output.

11. The primary goal of a future *smart grid* is to reliably maintain the balance of *active power* and *reactive power* while respecting the constraints of the grid, i.e., to provide for a stable supply of electric power in the face of volatile demand and supply.

12. The secondary goal of a future smart grid is to lower losses through an intelligent operation of the grid and its assets.

13. The power grid of today features increasingly more sensors that gather not only data on power flow, but also auxiliary data, such as weather conditions.

14. The notion of the smart grid does not only mean sensors, but also introduces *smart customers*, such as factories, cold storage houses, metal foundries, etc., that become subject to grid operator control.

15. The high volatility of power generation of renewable energy sources necessitates dynamic, node-local forecasting.

16. Given the sensory data, reliable, node-local forecasting, and controllable entities in the power grid, the problem of power demand and supply becomes a problem of information processing and distribution.

17. The amount of data from sensors and information coming from forecasts and analyses warrants a *divide-et-impera* approach to the smart grid.

18. Each important node in the power grid, from the traditional power plant, wind farm, PV power plant to the power substation, secondary customers, and domestic feed-in transformers, become *agents* that maintain their local power balance and act proactively to avoid a power disequilibrium.

19. A complex system such as a distributedly-controlled power grid requires simulation with external data sources providing data such as weather data and consumer behavior data.

20. External data sources may have varying degrees of resolution and accuracy; therefore, their quality must be automatically assessed by the simulation environment wherever possible.

21. Each agent features a modularized approach in which the main modules are the Forecaster Module, the Learner Module, the Messaging Module, and the Demand-Supply Module.

22. Each agent produces a local forecast of its local power balance using *Artificial Neural Networks* (ANN) that are trained from historical data.

23. The agent maintains a *pattern database* for supervised training of the agent-local ANN whose size is bounded by the dimensions of the ANN, maintaining a set of distinct (in contrast to all available) patterns.

24. The size of the ANN is adjusted in addition to the supervised training to mitigate worsening of the training error.

25. Weather and consumer data follow a time-series pattern; the agent's ANN therefore requires an internal representation of the concept of time, which makes a *Recurrent Neural Network* (RNN) suited for forecasting.

26. Effective training of an RNN is achieved by an evolutionary algorithm that includes the population's implicit gradient and dynamically controls the offspring's spread.

27. RNN-based forecasting achieves results reliable enough for grid operations planning, but cannot forecast extraordinary conditions the likes have never been encountered; it is therefore a sensible augmentation for meteorological forecasting.

28. Communication of power disequilibria and potential solutions between the agents occurs on an overlay network that models the power grid in communications architecture.

29. Agents are *pari passu*: Each agent can reply with a solution to the power disequilibrium and is a router of messages.

30. The impedance of sections of the power grid is the metric used to conduct routing in the agent communications overlay network.

31. No agent possesses a priori knowledge of other agents: Every node can potentially influence the electrical grid's power balance positively or negatively.

32. In the agent communication network, ad-hoc routing must be employed; selective broadcasting of requests and direct routing of replies minimizes the number of circulating messages while still reaching all agents.

33. Each agent must employ a message journal, noting messages passing through, sent, or answered by it, in order to establish ad-hoc routes to remote agents during the process of solving a power disequilibrium.

34. An agent maintains an internal power balance in which local as well as remote power disequilibria are noted as requirements when they concern the respective agent: Every node must first accommodate the power grid's global power balance.

35. Selecting replies by agents to form the solution to a power disequilibrium is a combinatorial problem that can be modelled in the Boolean domain.

36. All requirements are decomposed into atoms of equal size for representation in the Boolean domain: The responses are expressed through their acceptance function, the initial requirement through a symmetric function.

37. The representation of all requirements through *Ternary Vector Lists* (TVL) allows for an elegant formulation of the combinatorial problem and its solution.

38. The commutative nature of the intersection and the possibility to generate all permutations of a symmetric function reduces the space requirements of the corresponding TVLs to a minimum.

39. The utilization of TVLs and the XBOOLE system is more space-efficient than a representation with *Binary Decision Diagrams* (BDD) and more computation-efficient than any graph-based representation of the problem.

40. The agent-based approach is more time-efficient and space-efficient than a centralized, BDD-based approach in complex situations, and less efficient in simple situations due to its distributed nature.

41. Smart grid operation approaches should be measured against each other by two metrics introduced in this thesis: The effect of data in terms of power transmitted per byte of data volume required, and the data efficiency in terms of line loss avoided per byte of data volume required.

42. Distributed planning of demand and supply leads to an increased percentage of electrical energy from renewable energy sources.

43. Markets and price competition can be viewed as an optimization problem that arises after at least two possible solutions to a demand-supply calculation exist.

B   Protocol Message Types and Data Fields

B.1   Data and Field Types

ID fields

A 128 bit wide opaque identifier that is used to uniquely identify messages and nodes. At a given time, every ID string active in the network must be globally unique. How this unique ID is generated is up to the implementor. For agent IDs, an allocation scheme similar to that of MAC addresses would be appropriate. Message IDs can be randomly generated, for example, as UUIDs.

In *JavaScript Object Notation* (JSON) notation, an ID is represented by a string of printable characters; in binary notation, there is no constraint with regards to the identifier's encoding.

Please note that message IDs must be unique on their own; the uniqueness is not achieved through combining the information that the agent ID and that the message ID yields.

Message Type

An enumeration of the messages presented in Appendix B.2. The JSON encoding notes the message type as an unsigned integer. The binary representation is the binary encoding of the number given in the *message type* field as a field of 16 bit width.

Is Answer and Answer To

This field combination designates whether the current message is an answer to another particular message, or not.

The *is answer* field is a simple Boolean value. The JSON encoding variant of the *Lightweight Power Exchange Protocol* (LPEP) therefore uses the built-in JSON Boolean type. If the binary encoding scheme is used, the *is answer* field is 16 bit wide field, set to 16 0-bits for `false` and 16 1-bits for `true`.

The *answer to* field is an ID-type field that must follow immediately after the *is answer* field.

Time To Live (TTL)

Denotes the message's remaining *Time To Live* (TTL), an unsigned integer that is reduced by 1 with every forwarding. If the TTL value of a message reaches 0, it must not be forwarded.

The JSON encoding variant's key for this field is `ttl` and represents it as a JSON integer. The binary encoding for the TTL is a 32 bit wide field that encodes the unsigned integer in *network byte order*.

Timestamps

A point in time represented by the *Temps Atomique International (en. International Atomic Time)* (TAI) standard. More specifically, an external TAI64 label is used that represents a second using eight bytes in big endian notation.

The JSON notation represents a TAI64 external label as a string of hexadecimal characters. When using binary encoding, the bytes of the TAI64 label are transmitted directly as a 16 bit wide field.

Timespan

A timespan is defined as an interval of two TAI64 external labels. The interval boundaries are closed (left) and open (right). I.e., if t_1 and t_2 denote two TAI64 labels in TAI64 external format, a timespan is defined as $[t_1; t_2)$. A timestamp t is then part of the interval, if $t_1 \leq t < t_2$. Two time intervals, $[t_1; t_2)$ and $[t_3; t_4)$ will not overlap if $t_2 < t_3$. However, t_3 must follow immediately after t_2.

The JSON notation format of the LPEP represents a timespan as an array of two TAI64 external labels in string format. If a byte encoding scheme is used, the interval is represented by two concatenated timestamps.

Power Values, Power Value Intervals, and Power Types

A power value consists of a number and a unit denominator. The value is represented by a 32 bit unsigned integer, i.e., no fractions are permitted. In order to avoid rounding and cancellation errors that are common pitfalls to IEEE 753r floating-point numbers to propagate through the network, the protocol transmits only integers.

Power Intervals are closed intervals of two power values, i.e., $[P_1; P_2]$.

Permitted units are given by the following enumeration:

1. Kilowatt (`active`)

2. Kilovoltampere reactive (`reactive`).

The JSON encoding schema uses the `value` key for the number that uses the JSON built-in integer type and a string for the power type enumeration.

The binary representation of a power value is a concatenation of the 32 bit unsigned integer, transmitted in network order and the chosen value of the unit enumeration, encoded as a 16 bit field.

B.2   Message Types

Echo Request

Description

Requests an *echo reply* from another host. May only be used on immediately connected hosts; an agent must not forward an echo request. An echo request must be answered by an echo reply upon reception.

Data Fields

Name	JSON key	Value(s)
Message ID	id	The message's ID
Message Type	type	1
Sender ID	sender	The ID of the sender of the particular message
Is Answer	isAnswer	false
Answer To	answerTo	Not interpreted

Echo Reply

Description

The answer to an echo request, i.e., the 'pong' to a 'ping.' An agent must answer to an echo request with an echo reply and do so immediately. Echo replies must not be forwarded.

Data Fields

Name	JSON key	Value(s)
Message ID	`id`	The message's ID
Message Type	`type`	2
Sender ID	`sender`	The ID of the sender of the particular message
Is Answer	`isAnswer`	true
Answer To	`answerTo`	The ID of the echo request message this reply answers to

Online Notification

Description

The online notification informs other agents that a particular agent will come online (i.e., connected to the power grid) at a certain time. Online notifications serve to introduce a newly connected agent or announce the presence of one that was previously offline. Directly connected agents can thus learn the ID of their new neighbor before that neighbor is synchronized to the power grid. This helps to establish the communication infrastructure before an agent can exert its influence on the power grid. An agent must send an online notification message over all its connections after its software has booted and before it influences the power grid's power balance.

An online notification message must not be forwarded.

Data Fields

Name	JSON key	Value(s)
Message ID	id	The message's ID
Message Type	type	3
Sender ID	sender	The ID of the sender of the particular message
Is Answer	isAnswer	false
Answer To	answerTo	Not interpreted
Timestamp	timestamp	The TAI64 label denoting at which time the agent sending this message comes online

Offline Notification

Description

The offline notification message is the counterpart to the online notification message: It notifies the agent's neighbors that it will disconnect from the power grid at a certain time. An agent must send an offline notification message over all connection links before it loses synchronization with the power grid. Note that a disconnection from the power grid does not necessarily imply that the agent's software also shuts down or it loses its connection to the communication network.

An offline notification message must not be forwarded.

Sending an offline notification message with a timestamp that designates a point of time during which the agent still fulfills a request is a violation of correct protocol behavior. The agent must first ensure that the request is met and the grid's power balance remains intact.

Data Fields

Name	JSON key	Value(s)
Message ID	id	The message's ID
Message Type	type	4
Sender ID	sender	The ID of the sender of the particular message
Is Answer	isAnswer	false
Answer To	answerTo	Not interpreted
Timestamp	timestamp	The TAI64 label denoting at which time the agent sending this message goes offline

Demand Notification

Description

An agent sends a demand notification message when it requests additional power from other agents. Demand notification messages must be forwarded according to the directives defined in Section 6.2.

In addition to the timespan for which the request expressed by the demand notification message applies, the message also indicates the latest point at which an answer to it may arrive in order to be considered. Consequently, it is not necessary to answer such a request immediately, but within the time frame given by the *answer until* field.

The demand notification's *answer to* field can be either set to `true` or to `false`. If set to `true`, the demand notification message must be the answer to a previously received offer notification message. This accommodates the situation in which a surplus of power has been forecasted at another agent's node and is consequently answered by others. The reverse is true when the *is answer* field is set to `false`: Then, a demand for power has been forecasted in the agent's own locality that must be answered by other agents through offer notification messages.

A demand notification message that is an answer must eventually be followed by an acceptance notification by the original requester. Otherwise, it expires at the time indicated by the *answer until* field. If the agent does not receive an acceptance notification, it must not exert its influence on the power grid with regards to the original request received.

Data Fields

Name	JSON key	Value(s)
Message ID	`id`	The message's ID
Message Type	`type`	5
Sender ID	`sender`	The ID of the sender of the particular message
Receiver ID	`recever`	The ID of the receiver, if the message is an answer
Is Answer	`isAnswer`	`false`, if sent as the result of a forecast indicating an imbalance at the local node; `true` if intended as an answer to an offer notification

Answer To	answerTo	Only present if the message is an answer; if it is, this contains the original request's ID
TTL	ttl	The message's TTL
Distance	distance	The distance the message has travelled from sender to receiver
Timespan	timespan	The timespan interval for which this message is valid
Answer Until	answerUntil	A timestamp indicating when an answer can, at the latest, arrive at the requesting agent in order to be considered
Value	value	A power value interval denoting the amount requested
Type	powerType	The type of power requested

Offer Notification

Description

An agent expresses its ability (or desire) to deliver power to the grid with a message of this type. The same semantics apply as before to the demand notification message. The difference between the two message types is the flow of power: The demand notification message indicates a deficit at the sender's side, whereas the offer notification indicates a surplus of power at the sender's side.

Data Fields

Name	JSON key	Value(s)
Message ID	id	The message's ID
Message Type	type	6
Sender ID	sender	The ID of the sender of a particular message
Receiver ID	recever	The ID of the receiver, if the message is an answer
Is Answer	isAnswer	false, if sent as the result of a forecast indicating an imbalance at the local node; true if intended as an answer to a demand notifiction
Answer To	answerTo	Only present if the message is an answer; if it is, this contains the original request's ID
TTL	ttl	The message's TTL
Distance	distance	The distance the message has travelled from sender to receiver
Timespan	timespan	The timespan interval for which this message is valid
Answer Until	answerUntil	A timestamp indicating when an answer can, at the latest, arrive at the requesting agent in order to be considered

Value	`value`	A power value interval denoting the amount requested
Type	`powerType`	The type of power requested

Acceptance Notification

Description

An agent uses an acceptance notification message to inform another agent that it will take it up on its offer. Acceptance notification messages must be answers and an agent may formulate one only if an offer has been received. An offer whose *answer until* field contains a timestamp that has been reached or already lies in the past must not be answered with an acceptance notification message.

The acceptance notification also expresses the amount of power that is actually taken from the original offer. It must reside within the interval given in the original offer.

Data Fields

Name	JSON key	Value(s)
Message ID	id	The message's ID
Message Type	type	7
Sender ID	sender	The ID of the sender of a particular message
Receiver ID	recever	The ID of the receiver, if the message is an answer
Is Answer	isAnswer	true
Answer To	answerTo	The ID of the offer or demand notification this acceptance notification answers to
TTL	ttl	The message's TTL
Value	value	The power value accepted

Withdrawal Notification

Description

An agent may wish to withdraw an offer or demand notification it has sent at some time past. A case for this is described in Section 4.3; in short, an agent may wish to withdraw a request, demand notification and offer notification alike, when it receives a matching notification message of the opposite type that would solve its situation. It may also be possible that a forecast an agent has based a message on becomes invalid through a new calculation; it must then notify other agents that its message has become invalid by withdrawing it.

Withdrawal notification messages may be answers: If they are answers, they refer to a request an offer (or demand) notification message has been previously addressed. If the withdrawal notification message is not an answer, it must refer to an demand notification message or an offer notification message that originated at the current agent and was not formulated as an answer to a previously received request.

Data Fields

Name	JSON key	Value(s)
Message ID	`id`	The message's ID
Message Type	`type`	8
Sender ID	`sender`	The ID of the sender of a particular message
Receiver ID	`recever`	The ID of the receiver, if the message is an answer
Is Answer	`isAnswer`	`true` if the message's recipient has previously send a broadcast demand or offer notification; `false` if the agent revokes one of its own broadcast messages
Answer To	`answerTo`	The ID of the offer or demand notification this withdrawal addresses; can also address a non-answer demand or offer notification sent by the agent itself

| TTL | `ttl` | The message's TTL |

Acceptance Acknowledgement Notification

Description

The acceptance acknowledgement notification completes the four-way handshake and concludes the formation of a short-term contract. It must be an answer and may only be sent as an answer to an acceptance notification message. Consequently, it must be sent when an acceptance notification is received.

Data Fields

Name	JSON key	Value(s)
Message ID	id	The message's ID
Message Type	type	9
Sender ID	sender	The ID of the sender of the particular message
Receiver ID	recever	The ID of the receiver, if the message is an answer
Is Answer	isAnswer	true
Answer To	answerTo	The ID of the acceptance notification that is acknowledged by this message
TTL	ttl	The message's TTL

Constraint Notification

Description

Every agent is required to match or forward requests it receives. However, physical constraints can prohibit it from doing so. E.g., an agent can determine that if power would flow over a line represented by a candidate link for forwarding a request message, that link would be damaged. In such cases, the agent must replace the request with a constraint notification. A constraint notification must be forwarded. It contains the ID of the original request it replaced in a separate field. If an agent receives a constraint notification that matches a request it has received (or will receive), it may match or forward that request only if the distance value of the request is lower than that of the constraint message, indicating that power would flow over a different line or node than that which issued the constraint notification. If an agent receives a constraint notification to a request it has answered and the constraint notification's distance is equal or lower to that of the request, it must withdraw its response.

Data Fields

Name	JSON key	Value(s)
Message ID	id	The message's ID
Message Type	type	6
Sender ID	sender	The ID of the sender of a particular message
Receiver ID	recever	The ID of the receiver, if the message is an answer
Is Answer	isAnswer	false, if sent as the result of a forecast indicating an imbalance at the local node; true if intended as an answer to a demand notifiction
Answer To	answerTo	Only present if the message is an answer; if it is, this contains the original request's ID
TTL	ttl	The message's TTL
Distance	distance	The distance the message has travelled from sender to receiver

Constrained Message	`constrainedMessage`	
		ID of the original request the constraint message replaces

Bibliography

Abadi, M. and Cardelli, L. (1996). *A Theory of Objects*. Springer, Secaucus, NJ, USA, 1st edition.

Ackley, D. (1987). *A Connectionist Machine for Genetic Hillclimbing*, volume 28 of *The Kluwer International Series in Engineering and Computer Science*. Kluwer Academic Publishers, Boston, MA, USA.

AGEB (2015). Anteil Erneuerbarer Energien am Bruttostromverbrauch in Deutschland in den Jahren 1990 bis 2014. Online. http://de.statista.com/statistik/daten/studie/2142/umfrage/erneuerbare-energien-anteil-am-stromverbrauch/ [Retrieved: 2016-11-02].

Akers, S. B. (1978). Binary decision diagrams. *IEEE Transactions on Computers*, C-27(6):509–516.

Allelein, H.-J., Bollin, E., Oehler, H., and Schelling, U. (2010). *Energietechnik*. Studium: Energie und Umwelt. Vieweg+Teubner, Wiesbaden, Germany, 5th edition.

Anthony, M. and Bartlett, P. L. (2009). *Neural Network Learning: Theoretical Foundations*. Cambridge University Press, Cambridge, United Kingdom.

Aquino-Lugo, A. A., Klump, R., and Overbye, T. J. (2011). A control framework for the smart grid for voltage support using agent-based technologies. *IEEE Transactions on Smart Grid*, 2(1):173–180.

Aquino-Lugo, A. A. and Overbye, T. J. (2010). Agent technologies for control applications in the power grid. In *43rd Hawaii International Conference on System Sciences (HICSS)*, pages 1–10, Hawaii, USA.

Atkinson, R. (1996). IPv6 routing table size issues. Internet draft, Internet Engineering Taskforce (IETF). Online. `https://www.ietf.org/archive/id/draft-ietf-ipngwg-ipv6-routing-00.txt` [Retrieved: 2016-11-07].

Bäck, T. (1996). *Evolutionary Algorithms in Theory and Practice: Evolution Strategies, Evolutionary Programming, Genetic Algorithms*. Oxford University Press, New York, NY, USA.

Bajaj, S., Breslau, L., Estrin, D., Fall, K., Floyd, S., Haldar, P., Handley, M., Helmy, A., Heidemann, J., Huang, P., Kumar, S., McCanne, S., Rejaie, R., Sharma, P., Varadhan, K., Xu, Y., Yu, H., and Zappala, D. (1999). Improving simulation for network research. Technical Report 99-702b, University of Southern California, Los Angeles, CA, USA.

Banks, J., Carson, J., Nelson, B., and Nicol, D. (2013). *Discrete-Event System Simulation*. Pearson Education, Upper Saddle River, NJ, USA, 5th edition.

Barr, R., Haas, Z., and Renesse, R. V. (2004). JiST: Embedding simulation time into a virtual machine. In *5th EuroSim Congress on Modelling and Simulation*, Paris, France.

Baum, E. B. and Haussler, D. (1989). What size net gives valid generalization? *Neural Computation*, 1(1):151–160.

Beckmann, N., Kriegel, H.-P., Schneider, R., and Seeger, B. (1990). The R*-tree: An efficient and robust access method for points and rectangles. In *Proceedings of the 1990 ACM SIGMOD International Conference on Management of Data (SIGMOD '90)*, pages 322–331, New York, NY, USA. ACM.

Bellifemine, F. L., Caire, G., and Greenwood, D. (2007). *Developing Multi-Agent Systems with JADE*, volume 7 of *Wiley Series in Agent Technology*. John Wiley & Sons, Chichester, United Kingdom.

Bennett, C. H. (2003). Notes on Landauer's Principle, reversible computation, and Maxwell's Demon. *Studies in History and Philosophy of Science, Part B: Studies in History and Philosophy of Modern Physics*, 34(3):501–510.

Berg, H.-P. and Fritze, N. (2011). Reliability of main transformers. *Reliability: Theory and Applications*, 2(1):52–69.

Berndt, H., Hermann, M., Kreye, H. D., Reinisch, R., Scherer, U., and Vanzetta, J. (2007). TransmissionCode 2007 — Netz- und Systemregeln der deutschen Übertragungsnetzbetreiber. Online. `https://www.bdew.de/internet.nsf/id/` `A2A0475F2FAE8F44C12578300047C92F/$file/TransmissionCode2007.pdf` [Retrieved: 2016-11-01].

Bernstein, D. J. (1997). TAI64, TAI64N, and TAI64NA. Online. `http://cr.yp.to/libtai/tai64.html` [Retrieved: 2016-11-02].

Betz, A. (1926). *Windenergie und ihre Ausnutzung durch Windmühlen.* Vandenhoeck, Göttingen, Germany.

Bochmann, D. and Steinbach, B. (1991). *Logikentwurf mit XBOOLE.* Verlag Technik, Berlin, Germany, 1st edition.

Boole, G. (1847). *The Mathematical Analysis of Logic Being an Essay Towards a Calculus of Deductive Reasoning.* Philosophical Library, New York, NY, USA.

Booth, T. L. (1967). *Sequential Machines and Automata Theory.* John Wiley & Sons, New York, NY, USA.

Branke, J. (1995). Evolutionary algorithms for neural network design and training. In Alander, J. T., editor, *Proceedings of the First Nordic Workshop on Genetic Algorithms and its Applications (1NWGA)*, pages 145–163, Vaasa, Finland.

Brauner, G., Glaunsinger, W., Bofinger, S., John, M., Magin, W., Pyc, I., Schüler, S., Schulz, S., Schwing, U., Seydel, P., and Steinke, F. (2012). VDE-Studie: Erneuerbare Energie braucht flexible Kraftwerke — Szenarien bis 2020. Technical report, Energietechnische Gesellschaft im VDE (ETG), Frankfurt am Main, Germany.

Bray, T. (2014). The JavaScript Object Notation (JSON) data interchange format. RFC 7159, Internet Engineering Task Force (IETF).

Bray, T., Paoli, J., Sperberg-McQueen, C. M., Maler, E., and Yergeau, F. (1998). Extensible markup language (XML). World Wide Web Consortium Recommendation REC-xml-19980210 16, World Wide Web Consortium. Online. `http://www.w3.org/TR/1998/REC-xml-19980210` [Retrieved: 2016-11-01].

Brooks, R. J. and Tobias, A. M. (1996). Choosing the best model: Level of detail, complexity, and model performance. *Mathematical and Computer Modelling*, 24(4):1–14.

Brunner, C. (2008). IEC 61850 for power system communication. In *2008 IEEE/PES Transmission and Distribution Conference & Exposition*, pages 1–6, Chicago, IL, USA. IEEE.

Bryant, R. E. (1986). Graph-based algorithms for boolean function manipulation. *IEEE Transactions on Computers*, C-35(8):677–691.

Bryson, A. E. and Ho, Y.-C. (1969). *Applied Optimal Control: Optimization, Estimation and Control.* Hemisphere Publishing Corporation, Washington, DC, USA.

Buchholz, B. M., Bühner, V., Berninger, U., Fenn, B., and Styczynski, Z. A. (2012). Intelligentes Lastmanagement — Erfahrungen aus der Praxis. In *VDE-Kongress 2012*, Frankfurt am Main, Germany. VDE VERLAG GmbH.

Bundesnetzagentur (2015). Kraftwerksliste der Bundesnetzagentur — Stand 01.06.2015. Online. `http://www.bundesnetzagentur.de/cln_1431/DE/ Sachgebiete/ElektrizitaetundGas/Unternehmen_Institutionen/ Versorgungssicherheit/Erzeugungskapazitaeten/Kraftwerksliste/ kraftwerksliste-node.html` [Retrieved: 2015-09-16].

Bundesnetzagentur für Elektrizität, Gas, Telekommunikation, Post und Eisenbahnen (2014). Leitfaden zum EEG-Einspeisemanagement — Abschaltrangfolge, Berechnung von Entschädigungszahlungen und Auswirkungen auf die Netzentgelte. Version 2.1.

Burger, B., Kiefer, K., Kost, C., Nold, S., Philipps, S., Preu, R., Rentsch, J., Schlegl, T., Stryi-Hipp, G., Willeke, G., Wirth, H., Brucker, I., Häberle, A., and Warmuth, W. (2016). Photovoltaics report. Technical report, Fraunhofer ISE, Freiburg, Germany. Online. `https://www.ise.fraunhofer.de/de/downloads/pdf- files/aktuelles/photovoltaics-report-in-englischer-sprache.pdf` [Retrieved: 2016-11-01].

Bush, S. F. (2014). *Smart Grid — Communication-enabled Intelligence for the Electric Power Grid.* Wiley IEEE Series. John Wiley & Sons, Chichester, United Kingdom, 1st edition.

Butler, K. L., Sarma, N., and Prasad, V. R. (1999). A new method of network reconfiguration for service restoration in shipboard power systems. In *IEEE Transmission and Distribution Conference*, volume 2, pages 658–662, New Orleans, LA, USA. IEEE.

Calpe, C. (2015). DISCERN — distributed intelligence for cost-effective and reliable distribution network operation. *IEEE Smart Grid*. Online. `http://smartgrid.ieee.org/newsletters/july-2015/discern-distributed-intelligence-for-cost-effective-and-reliable-distribution-network-operation` [Retrieved: 2016-11-01].

Cannon, R. (2010). Potential impacts on communications from IPv4 exhaustion & IPv6 transition. Staff Working Paper 3, Federal Communications Commission, Washingtion, DC, USA.

Cao, L. C. L., Gorodetsky, V., and Mitkas, P. (2009). Agent mining: The synergy of agents and data mining. *IEEE Intelligent Systems*, 24(3):64–72.

Carlin, P. W., Laxson, A. S., and Muljadi, E. B. (2003). The history and state of the art of variable-speed wind turbine technology. *Wind Energy*, 6(2):129–159.

Chan, M. C. and Ramjee, R. (2005). TCP/IP performance over 3G wireless links with rate and delay variation. *Wireless Networks*, 11(1-2):81–97.

Chwif, L., Barretto, M. R. P., and Paul, R. J. (2000). On simulation model complexity. In *Proceedings of the 32nd conference on Winter Simulation*, pages 449–455, Orlando, FL, USA. Society for Computer Simulation International.

Clerc, M. (2012). Standard particle swarm optimisation. Technical report, HAL.

Cliff, D. (1997). Minimal-intelligence agents for bargaining behaviors in market-based environments. Technical Report September 1996, School of Cognitive and Computing Sciences, University of Sussex, Brighton, United Kingdom.

Coates, M. (2013). Revoking trust in two TurkTrust certificates. Technical report, Mozilla Cooperation. Online. `https://blog.mozilla.org/security/2013/01/03/revoking-trust-in-two-turktrust-certficates/` [Retrieved: 2016-11-02].

Coltun, R., Ferguson, D., Moy, J., and Lindem, A. (2008). OSPF for IPv6. RFC 5340, Internet Engineering Task Force (IETF).

Courtois, P.-J. (1985). On time and space decomposition of complex structures. *Communications of the ACM*, 28(6):590–603.

Cowie, J. H., Nicol, D. M., and Ogielski, A. T. (1999). Modeling the global internet. *Computing in Science & Engineering*, 1(1):30–38.

cppreference.com Contributors (2015). std::next_permutation. Online. `http://en.cppreference.com/w/cpp/algorithm/next_permutation` [Retrieved: 2016-11-02].

Cybenko, G. (1988). Continuous valued neural networks with two hidden layers are sufficient. Technical report, Department of Computer Science, Tufts University, Medford/Somerville, MA, USA.

Cybenko, G. (1989). Approximation by superpositions of a sigmoidal function. *Mathematics of Control, Signals and Systems*, 2(4):303–314.

Dantzig, T., Mazur, J., and Mazur, B. (2007). *Number: The Language of Science*. A Plume Book—The Masterpiece Science Edition. Plume.

Davidson, E. M., McArthur, S. D. J., McDonald, J. R., Cumming, T., and Watt, I. (2006). Applying multi-agent system technology in practice: Automated management and analysis of SCADA and digital fault recorder data. *IEEE Transactions on Power Systems*, 21(2):559–567.

Decker, B. L. (2000). Department of defense world geodetic system 1984 (WGS84). NIMA TR 8350.2, National Imagery and Mapping Agency, Bethesda, MD, USA.

Deering, S. and Hinden, R. (1998). Internet protocol, version 6 (IPv6) specification. RFC 2460, Internet Engineering Task Force (IETF).

dena — Deutsche Energie-Agentur (2013). Stromumwandlung durch Transformatoren. Online. `http://www.effiziente-energiesysteme.de/themen/intelligente-stromnetze/stromumwandlung.html` [Retrieved 2015-09-16].

Deutsche Presseagentur (2014). Batteriepark soll Stromschwankungen ausgleichen. *Freie Presse*. Date: 2014-09-15.

Dierks, T. and Rescorla, E. (2008). The transport layer security (TLS) protocol version 1.2. RFC 5246, Internet Engineering Task Force (IETF). Updated by RFCs 5746, 5878, 6176, 7465, 7507, 7568.

Dijkstra, E. W. (1959). A note on two problems in connexion with graphs. *Numerische Mathematik*, 1(1):269–271. doi:10.1007/BF01386390.

Dijkstra, E. W. (1982). On the role of scientific thought. In Dijkstra, E. W., editor, *Selected Writings on Computing: A Personal Perspective*, chapter 10, pages 60–66. Springer, New York, NY, USA.

Dresig, F. (1992). *Gruppierung: Theorie und Anwendung in der Logiksynthese*. Number 145 in Fortschritt-Berichte der VDI-Zeitschriften, Reihe Elektrotechnik, Elektronik. VDI-Verlag, Düsseldorf, Germany.

Droms, R. (1999). Automated configuration of TCP/IP with DHCP. *IEEE Internet Computing*, 3(4):45–53.

Ehrenberg, A. S. C. (1999). What we can get from graphs, and why. *Journal of Targeting, Measurement and Analysis for Marketing*, 8(2):113–134.

Elman, J. L. (1990). Finding structure in time. *Cognitive Science*, 14(2):179–211.

ENERCON GmbH (2011). Enercon Produktübersicht. Online. `http://www.enercon.de/p/downloads/ENERCON_P_D_web.pdf` [Retrieved: 2013-02-11].

ETSI (2012). Open smart grid protocol. ETSI Group Specification GS OSG 001, European Telecommunications Standards Institute, Sophia Antipolis, France.

Euler, E. (2001). The failures of the mars climate orbiter and mars polar lander—a perspective from the people involved. *Advances in Astronautical Sciences*, 107:635–655.

European Parliament, Council (2009). Directive 2009/28/EC of the European Parliament and of the Council of 23 April 2009 on the promotion of the use of energy from renewable sources and amending and subsequently repealing Directives 2001/77/EC and 2003/30/EC (Text with EEA relevance). Official Journal of the European Union. Date of effect: 2009-06-25.

Ewald, R. and Uhrmacher, A. M. (2012). Setting up simulation experiments with SESSL. In *Proceedings of the 2012 Winter Simulation Conference (WSC' 12)*, page 379, Berlin, Germany. Curran Associates.

Ewald, R. and Uhrmacher, A. M. (2014). SESSL: A domain-specific language for simulation experiments. *ACM Transactions on Modeling and Computer Simulation (TOMACS)*, 24(2):11.

Fall, K. R. and Stevens, W. R. (2012). *TCP/IP Illustrated: The Protocols*, volume 1. Addison-Wesley, Upper Saddle River, New Jersey, USA, 2nd edition.

FAZ.NET (2015). „Wir haben heute alle zusammen Geschichte geschrieben". *Frankfurter Allgemeine Zeitung*. Online. `http://www.faz.net/aktuell/wirtschaft/klimagipfel/weltklimavertrag-angenommen-wir-haben-heute-alle-zusammen-geschichte-geschrieben-13963330.html` [Retrieved: 2016-11-02].

Ferguson, N. and Schneier, B. (2003). *Practical Cryptography*. John Wiley & Sons, New York, NY, USA.

Fishman, G. S. (2013). *Discrete-Event Simulation: Modeling, Programming, and Analysis*. Springer Series in Operations Research. Springer Science & Business Media, New York, NY, USA.

Fleischhauer, J. and Nelles, R. (2007). Brand im Atomkraftwerk Krümmel: Willkommener Störfall. *Spiegel Online*. Online. `http://www.spiegel.de/jahreschronik/a-521391.html` [Retrieved: 2016-11-01].

Flosdorff, R. and Hilgarth, G. (2005). *Elektrische Energieverteilung*. Vieweg+Teubner, Stuttgart, Germany, 9th edition.

Fox-Penner, P. (2010). *Smart Power: Climate Change, the Smart Grid, and the Future of Electric Utilities*. Island Press, Washington, DC, USA.

Fruchterman, T. M. J. and Reingold, E. M. (1991). Graph drawing by force-directed placement. *Software-Practice and Experience*, 21:1129–1164.

Fujimoto, Y. (2006). Distribution test feeders. Online. `http://www.hayashilab.sci.waseda.ac.jp/RIANT/riant_test_feeder.html` [Retrieved: 2016-11-02].

Gamma, E., Helm, R., Johnson, R., and Vlissides, J. (1995a). *Design Patterns: Elements of Reusable Object-Oriented Software*, chapter 4.6, pages 205–206. Addison-Wesley Professional Computing Series. Addison-Wesley, Reading, MA, USA.

Gamma, E., Helm, R., Johnson, R., and Vlissides, J. (1995b). *Design Patterns: Elements of Reusable Object-Oriented Software*, chapter 3.3, pages 107–116. Addison-Wesley, Reading, MA, USA.

Ghosh, T. K. and Prelas, M. A. (2009). *Energy Resources and Systems*, volume 1, Fundamentals and Non-Renewable Resources. Springer Science+Business Media, Dordrecht, The Netherlands, 1st edition.

Goldstein, B., Hiriart, G., Bertani, R., Bromley, C., Gutiérrez-Negrín, L., Huenges, E., Muraoka, H., Ragnarsson, A., Tester, J., and Zui, V. (2011). Geothermal energy. In *IPCC Special Report on Renewable Energy Sources and Climate Change Mitigation*, chapter 4, pages 401–436. Cambridge University Press, Cambridge, United Kingdom and New York, NY, USA.

Greenwald, A. R. and Kephart, J. O. (1999). Shopbots and pricebots. In *16th Joint International Conference on Artificial Intelligence (JICAI)*, pages 506–511, Stockholm, Sweden. Morgan Kauffmann.

Ha, S., Rhee, I., and Xu, L. (2008). CUBIC: a new TCP-friendly high-speed TCP variant. *ACM SIGOPS Operating Systems Review*, 42(5):64–74.

Haberman, B. (2002). Allocation guidelines for IPv6 multicast addresses. RFC 3307, Internet Engineering Task Force (IETF).

Handley, M. (2006). Why the internet only just works. *BT Technology Journal*, 24(3):119–129.

Hebb, D. O. (2012). *The organization of behavior: A neuropsychological theory.* Routledge, Taylor & Francis Group, New York, NY, USA.

Henke, C., Siddiqui, A., and Khondoker, R. (2010). Network functional composition: State of the art. In *2010 Australasian Telecommunication Networks and Applications Conference (ATNAC 2010)*, pages 43–48, Auckland, New Zealand.

Heuck, K., Dettmann, K.-D., and Schulz, D. (2010). *Elektrische Energieversorgung: Erzeugung, Übertragung und Verteilung elektrischer*

Energie für Studium und Praxis. Studium: Elektrotechnik. Vieweg+Teubner Verlag/Springer Fachmedien, Wiesbaden, Germany, 8th edition.

Higgins, N., Vyatkin, V., Nair, N. K. C., and Schwarz, K. (2011). Distributed power system automation with IEC 61850, IEC 61499, and intelligent control. *IEEE Transactions on Systems, Man and Cybernetics Part C: Applications and Reviews*, 41(1):81–92.

Hochreiter, S. and Schmidhuber, J. (1997). Long short-term memory. *Neural Computation*, 9(8):1735–1780.

Holma, H. and Toskala, A. (2007). *HSDPA/HSUPA for UMTS: high speed radio access for mobile communications*. John Wiley & Sons, Chichester, United Kingdom.

Huntington, E. V. (1904). Sets of independent postulates for the algebra of logic. *Transactions of the American Mathematical Society*, 5(3):288–288.

Huntington, E. V. (1933a). Boolean algebra. a correction to "new sets of independent postulates for the algebra of logic with special reference to Whitehead and Russell's principia mathematica". *Transactions of the American Mathematical Society*, 35(2):557–558.

Huntington, E. V. (1933b). New sets of independent postulates for the algebra of logic with special reference to Whitehead and Russell's principia mathematica. *Proceedings of the National Academy of Sciences of the United States of America*, 35(1):274–304.

Hyyryläinen, J. and Jantunen, I. (2006). SSI protocol specification, version 1.2. Technical report, Nokia. Online. http://www.janding.fi/iiro/papers/ SSI%20protocol%20specification_12.pdf [Retrieved 2016-11-02].

IEEE Standards Association (2015). IEEE standards registration authority. Online. https://regauth.standards.ieee.org/standards-ra- web/pub/view.html [Retrieved: 2016-11-02].

Igel, C. and Hüsken, M. (2000). Improving the Rprop learning algorithm. In *Proceedings of the Second ICSC International Symposium on Neural Computation (NC 2000)*, pages 115–121, Berlin, Germany. ICSC Academic Press.

Igel, C. and Hüsken, M. (2003). Empirical evaluation of the improved Rprop learning algorithms. *Neurocomputing*, 50:105–123.

Imroz Sohel, M., Sellier, M., Brackney, L. J., and Krumdieck, S. (2009). Efficiency improvement for geothermal power generation to meet summer peak demand. *Energy Policy*, 37(9):3370–3376.

Innis, G. S. and Rexstad, E. (1983). Simulation model simplification techniques. *Simulation*, 41(1):7–15.

Inoue, T., Takano, K., Watanabe, T., Kawahara, J., Yoshinaka, R., Kishimoto, A., Tsuda, K., Minato, S.-I., and Hayashi, Y. (2014). Distribution loss minimization with guaranteed error bound. *IEEE Transactions on Smart Grid*, 5(1):102–111.

Institute of Electrical and Electronics Engineers (IEEE) (2012a). IEEE 802.11™: Wireless lans. Technical report, IEEE Standards Association, New York, NY, USA.

Institute of Electrical and Electronics Engineers (IEEE) (2012b). IEEE 802.3™-2012 — IEEE standard for ethernet. Technical report, IEEE Standards Association, New York, NY, USA.

International Atomic Energy Agency (2015). Power reactor information system. Online. `https://www.iaea.org/pris/` [Retrieved 2016-11-02].

International Standards Organization (ISO) (2004). Data elements and interchange formats — information interchange — representation of dates and times. ISO 8601:2004, International Organization for Standardization, Geneva, Switzerland.

International Standards Organization (ISO) (2005). IEC 61970. Technical report, International Organization for Standardization, Geneva, Switzerland.

International Standards Organization (ISO) (2012). Information technology — Control network protocol — Part 1: Protocol stack. Technical Report 14908-1:2012, International Organization for Standardization, Geneva, Switzerland.

Jackson, P., Hariskos, D., Wuerz, R., Kiowski, O., Bauer, A., Friedlmeier, T. M., and Powalla, M. (2014). Properties of Cu(In,Ga)Se$_2$ solar cells with new record efficiencies up to 21.7%. *physica status solidi rrl — rapid research letters*, 9(1):28–31.

Jaeger, T., Sailer, R., and Zhang, X. (2003). Analyzing integrity protection in the SELinux example policy. In *Proceedings of the 12th Conference on USENIX Security Symposium (SSYM'03)*, volume 12, Berkeley, CA, USA. USENIX Association.

Jordan, M. I. (1986). Serial order: A parallel distributed processing approach. *Advances in Connectionist Theory Speech*, 121(ICS-8604):471–495.

Jordan, M. I. (1995). Why the logistic function? A tutorial discussion on probabilities and neural networks. Computational Cognitive Science Technical Report 9503, Massachusetts Institute of Technology, Cambridge, MA, USA.

Jovanovic, P. and Neves, S. (2015). Dumb crypto in smart grids: Practical cryptanalysis of the open smart grid protocol. In *22nd International Workshop on Fast Software Encryption*, pages 297–316, Istanbul, Turkey.

Judd, J. S. (1990). *Neural network design and the complexity of learning*. MIT Press, Cambridge, MA, USA.

Kam, T., Villa, T., and Brayton, R. (1998). Multi-valued decision diagrams: theory and applications. *Multiple-Valued Logic*, 4(1):9–62.

Kelton, W. D. and Law, A. M. (2000). *Simulation modeling and analysis*. McGraw-Hill Series in Industrial Engineering and Management Science. McGraw-Hill Education, New York, NY, USA, 3rd edition.

Kennedy, J. and Eberhart, R. (1995). Particle swarm optimization. In *Proceedings of the 1995 IEEE International Conference on Neural Networks*, volume 3, pages 1942–1948, Perth, Australia. The University of Western Australia, IEEE.

Kent, S. and Seo, K. (2005). Security architecture for the internet protocol. RFC 4301, Internet Engineering Task Force (IETF). Updated by RFC 6040.

Kephart, J. O. (2002). Software agents and the route to the information economy. In *Proceedings of the National Academy of Sciences of the United States of America*, volume 99, pages 7207–7213, Washington, DC, USA.

Kephart, J. O., Hanson, J. E., and Greenwald, A. R. (2000). Dynamic pricing by software agents. *Computer Networks*, 32(6):731–752.

Kim, Y.-J., Kolesnikov, V., Kim, H., and Thottan, M. (2011). SSTP: a scalable and secure transport protocol for smart grid data collection. In *2011 IEEE International Conference on Smart Grid Communications (SmartGridComm)*, pages 161–166, Bruxelles, Belgium. IEEE.

Kirkpatrick, S., Gelatt, C. D., and Vecchi, M. P. (1983). Optimization by simulated annealing. *Science*, 220(4598):671–680.

Knaak, N. and Page, B. (2006). Applications and extensionds of the unified modelling language UML 2 for discrete event simulation. *Simulation*, 7(6):33–43.

Knott, C. G. (1911). *Quote from undated letter from Maxwell to Tait*, page 215. Cambridge University Press, Cambridge, United Kingdom.

Koritarov, V. S. (2004). Real-world market representation with agents. *IEEE Power and Energy Magazine*, 2(4):39–46.

Kreher, R. and Ruedebusch, T. (2007). *UMTS Signaling: UMTS Interfaces, Protocols, Message Flows and Procedures Analyzed and Explained*. John Wiley & Sons, Chichester, United Kingdom.

Kutter, I. and Rauner, M. (2012). „Das wäre ein Riesenproblem". *Die ZEIT*, (50).

Lai, Y.-T. (1993). *Logic verification and synthesis using function graphs*. PhD thesis, Computer Engineering, University of Southern California, Los Angeles, CA, USA.

Lalis, J. T., Gerardo, B. D., and Byun, Y.-C. (2014). An adaptive stopping criterion for backpropagation learning in feedforward neural network. *International Journal of Multimedia and Ubiquitous Energineering*, 9(8):149–156.

Landauer, R. (1961). Irreversibility and heat generation in the computing process. *IBM Journal of Research and Development*, 5(3):183–191.

Law, A. M. (1991). Simulation-models level of detail determines effectiveness. *Industrial Engineering*, 23(10):16.

Le Cun, Y., Denker, J. S., and Solla, S. A. (1990). Optimal brain damage. *Advances in Neural Information Processing Systems*, 2(1):598–605.

Leach, P., Mealling, M., and Salz, R. (2005). Universally unique identifier (UUID) URN namespace. RFC 4122, Internet Engineering Task Force (IETF). Online. `https://www.ietf.org/rfc/rfc4122.txt` [Retrieved: 2016-11-02].

Lee, Y., Durand, A., Woodyatt, J., and Droms, R. (2011). Dual-stack lite broadband deployments following IPv4 exhaustion. RFC 6333, Internet Engineering Task Force (IETF).

Liao, G.-C. and Tsao, T.-P. (2006). Application of a fuzzy neural network combined with a chaos genetic algorithm and simulated annealing to short-term load forecasting. *IEEE Transactions on Evolutionary Computation*, 10(3):330–340.

Liu, Z., Gao, W., Wan, Y.-H., and Muljadi, E. (2012). Wind power plant prediction by using neural networks. In *Proceedings of the 2012 IEEE Energy Conversion Congress and Exposition (ECCE)*, pages 3154–3160, Raleigh, NC, USA. IEEE.

Lobao, E. C. and Porto, A. J. V. (1997). A simulation study systematization. In *Proceedings of the XVII ENEGEP — National Congress of Industrial Engineering*, Gramado, Rio Grande do Sul, Brazil.

LoRa® Alliance (2016). LoRa® Technoloy. Online. `http://www.lora-alliance.org/What-Is-LoRa/Technology` [Retrieved: 2016-11-01].

Lysaght, P., Stockwood, J., Law, J., and Girma, D. (1994). Artificial neural network implementation on a fine-grained FPGA. In Hartenstein, R. W. and Servít, M. Z., editors, *Proceedings of the 4th International Workshop on Field-Programmable Logic and Applications (FPL'94)*, pages 421–431, Prague, Czech Republic. Springer.

MacDowell, J., Dutta, S., Richwine, A., Achilles, S., and Miller, N. (2015). Serving the future. *IEEE power & energy magazine*, 13(6):22–30.

Manwell, J. F., McGowan, J. G., and Rogers, A. L. (2010). *Wind Energy Explained: Theory, Design and Application*. John Wiley & Sons, Chichester, United Kingdom, 2nd edition.

Maqsood, I., Khan, M., and Abraham, A. (2004). An ensemble of neural networks for weather forecasting. *Neural Computing and Applications*, 13(2):112–122.

Marlinspike, M. (2009). Defeating OCSP with the character '3'. *Blackhat 2009*.

Maxwell, J. C. (2011). *Theory of Heat*. Cambridge University Press, Cambridge, United Kingdom.

McArthur, S. D. J., Davidson, E. M., Catterson, V. M., Dimeas, A. L., Hatziargyriou, N. D., Ponci, F., and Funabashi, T. (2007a). Multi-agent systems for power engineering applications—Part I: Concepts, approaches, and technical challenges. *IEEE Transactions on Power Systems*, 22(4):1743–1752.

McArthur, S. D. J., Davidson, E. M., Catterson, V. M., Dimeas, A. L., Hatziargyriou, N. D., Ponci, F., and Funabashi, T. (2007b). Multi-agent systems for power engineering applications—Part II: Technologies, standards, and tools for building multi-agent systems. *IEEE Transactions on Power Systems*, 22(4):1753–1759.

McCulloch, W. S. and Pitts, W. (1943). A logical calculus of the ideas immanent in nervous activity. *Bulletin of Mathematical Biophysics*, 5(4):115–133.

Merrion, P. (2011). Pilot test of ComEd's smart grid shows few consumers power down to save money. *Crain's Chicago Business*.

Minato, S.-I. (1993). Zero-suppressed BDDs for set manipulation in combinatorial problems. In *30th ACM/IEEE Design Automation Conference*, pages 272–277, Dallas, TX, USA. IEEE.

Momoh, J. (2012). *Smart Grid — Fundamentals of Design and Analysis*. IEEE Press Series on Power Energineering. IEEE Press, Piscataway, NJ, USA.

Mozer, M. C. (1989). A focused backpropagation algorithm for temporal pattern recognition. *Complex Systems*, 3(4):349–381.

Nagata, T. and Sasaki, H. (2002). A multi-agent approach to power system restoration. *IEEE Transactions on Power Systems*, 17(2):457–462.

Nagayama, S. and Sasao, T. (2007). Representations of elementary functions using edge-valued MDDs. In *Proceedings of The 37th International Symposium on Multiple-Valued Logic (ISMVL 2007)*, Oslo, Norway. IEEE.

Nara, K., Shiose, A., Kitagawa, M., and Ishihara, T. (1992). Implementation of genetic algorithm for distribution systems loss minimum re-configuration. *IEEE Transactions on Power Systems*, 7(3):1044–1051.

Nationaal Cyber Security Centrum (2011). Frauduleus uitgegeven beveiligingscertificaat ontdekt. Online. `https://www.ncsc.nl/actueel/factsheets/factsheet-frauduleus-uitgegeven-certificaat-ontdekt.html` [Retrieved: 2016-11-02].

National Museum of American History (1919). Powering a generation of change. Online. `http://americanhistory.si.edu/powering/basics/load.htm` [Retrieved: 2016-11-01].

National Security Agency (2013). Security-enhanced linux. Online. `http://www.selinuxproject.org/` [Retrieved: 2016-11-02].

Nurseitov, N., Paulson, M., Reynolds, R., and Izurieta, C. (2009). Comparison of JSON and XML data interchange formats: A case study. *Caine*, 2009:157–162.

Nwana, H. S. (1996). Software agents: An overview. *Knowledge Engineering Review*, 11(3):205–244.

Observ'ER, editor (2013). *Fifteenth Inventory*, chapter 3, pages 3–7. Observ'ER, 146, rue de l'Université, Paris, France, 2013 edition.

Oeding, D. and Oswald, B. R. (2011). *Elektrische Kraftwerke und Netze*. Springer, Berlin, Germany, 7th edition.

O'Malley, S. W. and Peterson, L. L. (1992). A dynamic network architecture. *ACM Transactions on Computer Systems*, 10(2):110–143.

Omondi, A. R. and Rajapakse, J. C., editors (2006). *FPGA Implementations of Neural Networks*, volume 365. Springer, Dordrecht, The Netherlands.

OPNET Technologies, Inc. (2015). OPNET Modeler. Online. `http://www.opnet.com/` [Retrieved: 2016-11-01].

Oswald, B. R. (2007). Verlust- und Verlustenergieabschätzung für das 380-kV-Leitungsbauvorhaben Wahle–Mecklar in der Ausführung als Freileitung oder Drehstromkabelsystem. Technical report, Universität Hannover, Hannover, Germany. Online. `http://www.netzausbau-niedersachsen.de/downloads/verlustvergleichwahlemecklarfinalv2.pdf` [Retrieved: 2016-11-01].

Padovan, B., Sackmann, S., Eymann, T., and Pippow, I. (2002). A prototype for an agent-based secure electronic marketplace including reputation tracking mechanisms. *International Journal of Electronic Commerce*, 6(4):93–113.

Peano, G. (1888). *Calcolo Geometrico*. Fratelli Bocca, Turin, Italy.

Pidd, M. (1999). Just modeling through: A rough guide to modeling. *Interfaces*, 29(2):118–132.

Pipattanasomporn, M., Feroze, H., and Rahman, S. (2009). Multi-agent systems in a distributed smart grid: Design and implementation. In *Power Systems Conference and Exposition, 2009. PSCE'09. IEEE/PES*, pages 1–8. IEEE.

Pöller, M. and Achilles, S. (2003). Aggregated wind park models for analyzing power system dynamics. In *Proceedings of the 4th International Workshop on Large-Scale Integration of Wind Power and Transmission Networks for Offshore Wind Farms*, pages 1–10, Billund, Denmark.

Postel, J. (1980). User datagram protocol. RFC 768, Internet Engineering Task Force (IETF).

Postel, J. (1981a). Internet protocol. RFC 791, Defense Advanced Research Projects Agency.

Postel, J. (1981b). Transmission control protocol. RFC 793, University of Southern California, Marina del Rey, CA, USA. Updated by RFCs 1122, 3168, 6093, 6528.

Posthoff, C. and Steinbach, B. (1979a). *Binäre dynamische Systeme — Algorithmen und Programme*, volume 8. Technische Hochschule Karl Marx Stadt, Karl-Marx-Stadt, German Democratic Republic.

Posthoff, C. and Steinbach, B. (1979b). *Binäre Gleichungen — Algorithmen und Programme*, volume 1 of *Wissenschaftliche Schriftenreihe der Technischen Hochschule Karl-Marx-Stadt*. Technische Hochschule Karl-Marx-Stadt, Karl-Marx-Stadt, German Democratic Republic.

Posthoff, C. and Steinbach, B. (2004). *Logic Functions and Equations: Binary Models for Computer Science*, chapter 9, pages 377–384. Springer, Dordrecht, the Netherlands.

Posthoff, C. and Steinbach, B. (2014). Solving the game of Sudoku. *ICGA Journal*, 37(2):111–116.

Powell, L. (2005). *Power System Load Flow Analysis*. McGraw-Hill Professional Engineering. McGraw-Hill, New York, NY, USA.

Reade, C. (1989). *Elements of Functional Programming*. International Computer Science Series. Addison-Wesley, Wokingham, United Kingdom.

Redl, S., Weber, M., and Oliphant, M. (1998). *GSM and Personal Communications Handbook*. Artech House Mobile Communications Series. Artech House, Boston, MA, USA.

Rekhter, Y. and Li, T. (1995). A border gateway protocol 4 (BGP-4). RFC 4271, Internet Engineering Task Force (IETF). Updated by RFCs 6286, 6608, 6793, 7606, 7607, 7705.

Reuther, B. and Henrici, D. (2008). A model for service-oriented communication systems. *Journal of Systems Architecture*, 54(6):594–606.

Richter, H. and Marz, L. (2000). Toward a standard process: the use of UML for designing simulation models. In *Proceedings of the 32nd Conference on Winter Simulation*, pages 394–398, Orlando, FL, USA. Society for Computer Simulation International.

Riedmiller, M. (1994a). Advanced supervised learning in multi-layer perceptrons — from backpropagation to adaptive learning algorithms. *Computer Standards & Interfaces*, 16(3):265–278.

Riedmiller, M. (1994b). Rprop — description and implementation details. Technical report, Institut für Logik, Komplexität und Deduktionssysteme, University of Karlsruhe, Karlsruhe, Germany.

Riedmiller, M. and Braun, H. (1992). RPROP — a fast adaptive learning algorithm. In *Proceedings of the International Symposium on Computer and Information Science VII*, Dallas, TX, USA.

Robinson, S. (1994). Simulation projects: Building the right conceptual model. *Industrial Engineering-Norcross*, 26(9):34–36.

Robinson, S. (2004). *Simulation — The Practice of Model Development and Use*. John Wiley & Sons, Chichester, United Kingdom.

Rogers, K. M., Klump, R., Khurana, H., Aquino-Lugo, A. a., and Overbye, T. J. (2010). An authenticated control framework for distributed voltage support on the smart grid. *IEEE Transactions on Smart Grid*, 1(1):40–47.

Rosenblatt, F. (1957). The perceptron: A perceiving and recognizing automaton. Technical Report 85-460-1, Project PARA, Cornell Aeronautical Laboratory, Ithaca, NY, USA.

Rumelhart, D. E., Hinton, G. E., and Williams, R. J. (1986). Learning representations by back-propagating errors. *Nature*, 323(6088):533–536.

Ruppert, M., Veith, E. M., and Steinbach, B. (2014). An evolutionary training algorithm for artificial neural networks with dynamic offspring spread and implicit gradient information. In *Proceedings of the Sixth International Conference on Emerging Network Intelligence (EMERGING 2014)*, Rome, Italy. International Academy, Research, and Industry Association.

Russel, S. and Norvig, P. (2010). *Artificial Intelligence — A Modern Approach*. Prentice Hall Series in Artificial Intelligence. Pearson Education, Boston, MA, USA, 3rd edition.

Rybach, L. (2007). Geothermal sustainability assessment framework. *Geo-Heat Centre Quarterly Bulletin*, 29(September):7.

Salt, J. D. (1993). Simulation should be easy and fun! In *Proceedings of the 25th conference on Winter simulation*, pages 1–5, Los Angeles, CA, USA. ACM.

Santos, O. (2014). Global internet routing table reaches 512k milestone. Online. `http://blogs.cisco.com/sp/global-internet-routing-table-reaches-512k-milestone` [Retrieved: 2016-11-02].

Savola, P. (2011). Overview of the internet multicast addressing architecture. RFC 6308, Internet Engineering Task Force (IETF).

Scalable Network Technologies (2016). Qualnet homepage. Online. `http://web.scalable-networks.com/content/qualnet` [Retrieved: 2016-11-02].

Schultze-Melling, J. (2010). Directive 2006/24/EC (data rentention directive). In Büllesbach, A., Gijrath, S., Poullet, Y., and Prins, C., editors, *Concise European IT Law*. Kluwer Law International, Alphen aan den Rijn, The Netherlands, 2nd edition.

Sculley, D., Holt, G., Golovin, D., Davydov, E., Phillips, T., Ebner, D., Chaudhary, V., and Young, M. (2014). Machine learning: The high interest credit card of technical debt. In *SE4ML: Software Engineering for Machine Learning (NIPS 2014 Workshop)*, pages 1–9, Montréal, Canada.

Sehgal, A., Perelman, V., Kuryla, S., and Schonwalder, J. (2012). Management of resource constrained devices in the internet of things. *IEEE Communications Magazine*, 50(12):144–149.

Seller, H. and Röderer, H. (2015). Stromausfall nach Explosion: Technischer Defekt war Ursache. *Badische Zeitung*. Published 2015-07-14. Online. `http://www.badische-zeitung.de/ortenaukreis/stromausfall-nach-explosion-technischer-defekt-war-ursache--107690745.html` [Retrieved: 2016-11-07].

Sesia, S., Toufik, I., and Baker, M. (2009). *LTE: the UMTS Long Term Evolution: From Theory to Practice*. John Wiley & Sons, Chichester, United Kingdom, 2nd edition.

Shannon, C. E. (1948). A mathematical theory of communication. *Bell System Technical Journal*, 27:379–423.

Shannon, C. E. (1959). Coding theorems for a discrete source with a fidelity criterion. In *International Convention Record*, volume 4, pages 142–163. Institute of Radio Engineers.

Shannon, R. E. (1998). Introduction to the art and science of simulation. In Medeiros, D. J., Watson, E. F., Carson, J. S., and Manivannan, M. S., editors, *Proceedings of the 30th Conference on Winter Simulation*, volume 1, pages 7–14. ACM.

Shi, Y. and Eberhart, R. (1998). A modified particle swarm optimizer. In *Proceedings of IEEE International Conference on Evolutionary Computation*, pages 69–73. IEEE.

Sietsma, J. and Dow, R. (1988). Neural net pruning—why and how. In *IEEE International Conference on Neural Networks*, pages 325–333, San Diego, CA, USA.

Skinner, B. F. (1953). *Science and Human Behavior*. Macmillan, New York, NY, USA, 1st edition.

Slootweg, J. G., Haan, S. W. H. D., Polinder, H., and Kling, W. L. (2002). Aggregated modelling of wind parks with variable speed wind turbines in power system dynamics simulations. In *14th Power Tech Conference Proceedings*, pages 24–28, Sevilla, Spain. IEEE.

Smith, R. G. (1980). The Contract Net Protocol: High-level communication and control in a distributed problem solver. *IEEE Transactions on Computer*, C-29(12):1104–1113.

Sobeih, A., Hou, J. C., Kung, L. C., Li, N., Zhang, H., Chen, W. P., Tyan, H. Y., and Lim, H. (2006). J-Sim: A simulation and emulation environment for wireless sensor networks. *IEEE Wireless Communications*, 13(4):104–119.

Stallings, W. (2013). *Cryptography and Network Security: Principles and Practice*. Prentice Hall, Upper Saddle River, NJ, USA, 6th edition.

Steinbach, B. (1984). *Theorie, Algorithmen und Programme für den rechnergestützten logischen Entwurf digitaler Systeme*. Dissertation B, Technische Hochschule Karl-Marx-Stadt, Karl-Marx-Stadt, German Democratic Republic.

Steinbach, B. (1992). XBOOLE — A toolbox for modeling, simulation, and analysis of large digital systems. *System Analysis and Modelling Simulation*, 9:297–312.

Steinbach, B. and Posthoff, C. (2012). Solutions of exceptionally complex boolean problems. In Steinbach, B., editor, *Proceedings of the 10th International Workshop on Boolean Problems*, pages 185–223, Freiberg, Germany. Verlag der Technischen Universität Bergakademie Freiberg.

Steinbach, B. and Posthoff, C. (2014). Four-colored rectangle-free grids: Four-colored rectangle-free grids of the size 12×21. In Steinbach, B., editor, *Recent Progress in the Boolean Domain*, pages 121 144. Cambridge Scholars Publishing, Newcastle upon Tyne, United Kingdom.

Steinbach, B. and Werner, M. (2014). XBOOLE-CUDA fast boolean operations on the GPU. In Steinbach, B., editor, *Proceedings of the 11th International Workshop on Boolean Problems*, pages 75–84, Freiberg, Germany. Verlag der Technischen Universität Bergakademie Freiberg.

Sterner, M. and Stadler, I. (2014). *Energiespeicher*. Springer, Berlin, Germany, 1st edition.

Stoica, I., Morris, R., Liben-Nowell, D., Karger, D. R., Kaashoek, M. F., Dabek, F., and Balakrishnan, H. (2003). Chord: a scalable peer-to-peer lookup protocol for internet applications. *IEEE/ACM Transactions on Networking*, 11(1):17–32.

Sverdlik, Y. (2014). BGP routing table size limit blamed for tuesday's website outages. Online.
`http://www.datacenterknowledge.com/archives/2014/08/13/bgp-routing-table-size-limit-blamed-for-tuesdays-website-outages/`
[Retrieved: 2016-11-02].

Tanenbaum, A. S. (2003). *Computer Networks*. Prentice Hall, Upper Saddle River, NJ, USA, 4th edition.

The NS-3 Project (2015). NS-3 Homepage. Online. `http://www.nsnam.org/`
[Retrieved: 2016-11-01].

Tsang, Y., Coates, M., and Nowak, R. D. (2003). Network delay tomography. *IEEE Transactions on Signal Processing*, 51(8):2125–2136.

Turing, A. M. (1950). Computing machinery and intelligence. *Mind*, pages 433–460.

US-Canada Power System Outage Task Force (2004). Final report on the August 14, 2003 blackout in the United States and Canada: Causes and recommendations. Technical report, Office of Electricity Delivery & Energy Reliability.

U.S. Department of Transportation, Federal Highway Administration, Office of Operations (2013). Simplified guide to the incident command system for transportation professionals. Online.
`http://ops.fhwa.dot.gov/publications/ics_guide/ics_guide.pdf`
[Retrieved: 2016-11-07].

Vale, Z., Pinto, T., Praça, I., and Morais, H. (2011). MASCEM: Electricity markets simulation with strategic agents. *IEEE Intelligent Systems*, 26(2):9–17.

Varga, A. (2001). The OMNeT++ discrete event simulation system. In *Proceedings of the 15th European Simulation Multiconference*, pages 319–324, Prague, Czech Republic.

Varga, A. and Hornig, R. (2008). An overview of the OMNeT++ simulation environment. In *Proceedings of the 1ˢᵗ International Conference on Simulation Tools and Techniques for Communications, Networks and Systems & Workshops*, page 60. Institute for Computer Sciences, Social-Informatics and Telecommunications Engineering (ICST), ACM.

VASCO Data Security International, Inc. (2011). DigiNotar reports security incident. Online.
`https://www.vasco.com/company/about_vasco/press_room/`
`news_archive/2011/news_diginotar_reports_security_incident.aspx`
[Retrieved: 2016-11-01].

VDMA Power Systems (2013). Fähigkeiten von Stromerzeugungsanlagen im Energiemix. Technical report, Verband Deutscher Maschinen- und Anlagenbau e.V. (VDMA), Frankfurt am Main, Germany.

Veith, E., Steinbach, B., and Windeln, J. (2014). A lightweight distributed software agent for automatic demand-supply calculation in smart grids. *International Journal On Advances in Internet Technology*, 7:97–113.

Veith, E. M. and Steinbach, B. (2015). Modeling demand and supply in a smart grid. In *Proceedings of the 24ᵗʰ International Workshop on Post-Binary ULSI Systems*, number May, pages 1–2, Waterloo, Canada. University of Waterloo.

Veith, E. M., Steinbach, B., and Windeln, J. (2013). A lightweight messaging protocol for smart grids. In *Proceedings of the Fifth International Conference on Emerging Network Intelligence (EMERGING 2013)*, pages 6–12, Porto, Portugal. IARIA XPS Press.

Vestas Wind Systems A/S (2012). 2 MW Platform. Online.
`https://www.vestas.com/en/products/turbines/v110%202_0_mw#!2mw-`
`platform` [Retrieved: 2016-11-02].

Vrudhula, S. B., Pedram, M., and Lai, Y.-T. (1996). Edge valued binary decision diagrams. In Sasao, T. and Fujita, M., editors, *Representations of Discrete Functions*, pages 109–132. Kluwer Academic, Boston, MA, USA.

Vyatkin, V., Zhabelova, G., Higgins, N., Schwarz, K., and Nair, N.-K. C. (2010a). Towards intelligent smart grid devices with IEC 61850 interoperability and IEC 61499 open control architecture. In *Transmission and Distribution Conference and Exposition 2010 IEEE PES*, volume 2, pages 1–8, Porto, Portugal. Instituto de Patologia e Imunologia Molecular da Universidade do Porto (IPATIMUP), Faculdade de Ciências da Universidade do Porto, Portugal, IEEE.

Vyatkin, V., Zhabelova, G., Higgins, N., Ulieru, M., Schwarz, K., and Nair, N.-K. C. (2010b). Standards-enabled smart grid for the future energy web. In *Proceedings of the 1ˢᵗ Conference on Innovative Smart Grid Technologies (ISGT)*, pages 1–9, Gothenburg, Sweden. IEEE.

Walling, R. and Shattuck, G. B. (2007). Distribution transformer thermal behavior and aging in local-delivery distribution systems. In *Proceedings of the 19ᵗʰ Conference on Electricity Distribution*, Vienna, Austria.

Wang, L., Chen, C., and Shen, T. (2014). Improvement of power flow calculation with optimization factor based on current injection method. *Discrete Dynamics in Nature and Society*, 2014.

Wang, Y. and Vassileva, J. (2003). Trust and reputation model in peer-to-peer networks. In *Proceedings of the Third International Conference on Peer-to-Peer Computing (P2P 2003)*, pages 150–157, Linköping, Sweden. IEEE.

Ward, S. C. (1989). Arguments for constructively simple models. *Journal of the operational research society*, 40(2):141–153.

Willemain, T. R. (1994). Insights on modeling from a dozen experts. *Operations Research*, 42(2):213–222.

Wooldridge, M. and Ciancarini, P. (2001). Agent-oriented software engineering: The state of the art. In *Agent-Oriented Software Engineering*, pages 1–28. Springer-Verlag, Heidelberg, Germany.

Xiong, L. and Liu, L. (2003). A reputation-based trust model for peer-to-peer e-commerce communities. In *IEEE International Conference on E-Commerce (CEC 2003)*, pages 275–284, Newport Beach, CA, USA. IEEE.

Zervos, A., Lins, C., and Muth, J. (2010). RE-thinking 2050 — a 100% renewable energy vision for the European Union. Online. `http://ec.europa.eu/clima/consultations/docs/0005/registered/91650013720-46_european_renewable_energy_council_en.pdf` [Retrieved: 2016-11-02].

Zhabelova, G. and Vyatkin, V. (2011). Multi-agent smart grid automation architecture based on IEC 61850/61499 intelligent logical nodes. *IEEE Transactions on Industrial Electronics*, 59(5):2351–2362.

Zhu, J. and Sutton, P. (2003). FPGA implementations of neural networks—a survey of a decade of progress. In *International Conference on Field Programmable Logic and Applications*, pages 1062–1066, Lisbon, Portugal. Springer.

Zimmermann, H. (1980). OSI reference model—the ISO model of architecture for open systems interconnection. *IEEE Transactions on Communications*, COM-28(4):425–432.

Zitterbart, M., Stiller, B., and Tantawy, A. N. (1993). Model for flexible high-performance communication subsystems. *IEEE Journal on Selected Areas in Communications*, 11(4):507–518.

Zyp, K., SitePen (USA), and Court, G. (2013a). JSON Schema: Core definitions and terminology. draft-zyp-json-schema 04, Internet Engineering Task Force (IETF).

Zyp, K., SitePen (USA), and Court, G. (2013b). JSON Schema: Interactive and non-interactive validation. draft-fge-json-schema-validation 00, Internet Engineering Task Force (IETF).